OPPENHEIMER

PORTRAIT OF AN ENIGMA

Jeremy Bernstein

Ivan R. Dee

CHICAGO 2004

Photo credits: frontispiece, CORBIS; page 2, J. Robert Oppenheimer Memorial Committee; page 30, Molly B. Lawrence, courtesy of AIP Emilio Segrè Visual Archives; page 64, Bettmann / CORBIS; page 92, *Herblock's Here and Now* (Simon & Schuster, 1955); page 170, Ralph Morse / Time Life Pictures / Getty Images; page 198, Randal Hagadorn.

Library of Congress Cataloging-in-Publication Data:
Bernstein, Jeremy, 1929–
    Oppenheimer : portrait of an enigma / Jeremy Bernstein.
      p. cm.
    Includes bibliographical references.
    ISBN 1-56663-569-1 (acid-free paper)
    1. Oppenheimer, J. Robert, 1904–1967. 2. Physicists—United States—Biography. I. Title.

QC16.O62B43 2004
530'.092—dc22
[B]   2003066652

PREFACE

IN THE 1960s, when I began writing profiles of scientists for *The New Yorker*, an obvious subject would have been Robert Oppenheimer. In the first place, I knew and admired him. During the two years I spent—1957–1959—at the Institute for Advanced Study in Princeton, where he was the director, I was able to observe him on an almost daily basis. We also had some private talks, one of the most memorable of them on a train ride from Princeton to New York, during which he spoke a little about what he referred to as his "case"—the famous 1954 hearing before the Atomic Energy Commission in which he lost his security clearance. This aside, many of my teachers had either been at Los Alamos or before the war had studied under Oppenheimer in California. They told me all sorts of stories.

I also knew something about his physics. There is a tendency to dismiss Oppenheimer as a physicist, perhaps partly because he seemed disappointed in what he had been able to do. But his standards were impossibly high. He was a superb physicist. Had he lived longer—he died in 1967, at the age of sixty-two—the pioneering work that he and his students did on neutron stars and black holes would have, I am convinced, been widely honored, possibly by a Nobel Prize. Because of his ability in physics, and his astounding charisma, he did at Los Alamos what I think no one else would have been able to do. If Oppenheimer had not been the director at Los Alamos, I am persuaded that, for better or worse, the Second World War would have ended very differently—without the use of nuclear weapons. They would not have been ready. In short, Oppenheimer was a profile writer's dream. Nonetheless I decided that he was not a subject for me.

There were several reasons, two of them involving *The New Yorker*. About this time William Shawn, its editor, told me that he had asked Oppenheimer to write something for the magazine. Oppenheimer was willing, but only on the condition that nothing he wrote could be changed in any way, including punctuation. Mr. Shawn explained to me that he could not possibly turn over his editorial responsibilities to Oppenheimer or anyone else. That ended that.

The second incident involved me directly. On November 18, 1962, Niels Bohr died at the age of seventy-seven. In the hierarchy of twentieth-century physicists I would place Bohr just below Einstein. I composed a tribute to him for *The New Yorker*, which was designed to go into the Notes and Comments section of the Talk of the Town. I wrote what I thought was an affectionate, brief profile-obituary. In it I quoted something that Oppenheimer had written about Bohr's contribution to the quantum theory. As was the custom of *The New Yorker*'s fact checkers, they called Oppenheimer. Not only did they vet the quote, they read him the whole piece as context. They then called me to report that Oppenheimer had told them that what I had written was terrible and that I should call him. When I did so, he more or less repeated this to me but offered no suggestions as to how I might improve the piece except to insist that I remove his quotation. Needless to say, I was crushed.

As it happened, I was then spending a fair amount of time at the Rockefeller University, where one of the senior professors was the late George Uhlenbeck. Uhlenbeck, who had been born in Java and schooled in Holland, was well known in our community as both a physicist and a person of great wisdom and common sense. He had known both Oppenheimer and Bohr for decades. Oppenheimer had spent time in Leiden and had learned Dutch. I decided I would show what I had written to Uhlenbeck, and if he too said it was awful I would withdraw it. Uhlenbeck read it and told me it was fine. He conjectured that Oppenheimer was annoyed because *he* had not been asked to write Bohr's obituary. Uhlenbeck also supplied the lyrics to a song they used to sing in Holland to the tune of the Dutch national anthem. Each verse ended with *"En Bohr, en Einstein hebben wij atijd vereed"* (And Bohr and Einstein we have always revered). This replaced the Oppenheimer quote in my obituary for Bohr. Incidentally, after Oppenheimer died I wrote *his* obituary for *The New Yorker*.

But there was an even deeper reason why I did not want to try to write Oppenheimer's profile. The way I worked was to spend hours taping interviews with my subjects. These recorded interviews really are oral biographies. If I was to do Oppenheimer, that is what I would have attempted. But even assuming he would grant me the time and access, I did not think I could ask him the questions I wanted to. I probably could have asked him how he felt about nuclear weapons and his role in them. Did he still feel, as he told President Truman, that he had "blood on his hands"? I am sure I could have discussed his physics and the physics of his contemporaries. But what about his personal life? Could I have asked him about his affair with Jean Tatlock in the late 1930s which led to his flirtation with the Communist party? Could I have asked him about his ambiguous feelings about being Jewish? Is this what caused him to drop his first name "Julius" for the letter "J," which he always insisted stood for "nothing"? I might have asked him why he never added anything to the art he inherited from his father—the van Goghs, among other things. These were the kinds of things one would wonder about if one was writing a biography as opposed to a hagiography. In short, I felt I would run into a wall, so I never got started.

Nonetheless the subject pursued me. Inevitably it came up in profiles of other scientists. I read and reread the transcript of Oppenheimer's hearing—a sort of tragic play—and any of the biographies that mentioned him. Facts kept accumulating like bales of hay. On the wall of the room in which I am writing there is a photograph of Oppenheimer. It is one that he gave my father. Oppenheimer is clearly dressed in one of the specially tailored suits he bought at Langrock's in Princeton. (They too had a picture of him.) His hair is cut very short, and his right ear protrudes. He looks both extremely intelligent and very wary. He is not looking at the photographer, and since the photo is in black and white you cannot see the remarkable blue of his eyes. He has a cigarette— they eventually killed him—in his left hand held between two tapered fingers. It is a face that would get your attention anywhere, even if you did not know anything about the person it belonged to. Someone once said that when Dr. Johnson entered a room he changed its electric charge. This was true of Oppenheimer. You could not be neutral.

As I accumulated my facts I kept staring at this picture and wondering if I now had enough distance from the events to write the profile I had meant to write in the 1960s. Things are both easier and more difficult—for the same reason. Nearly all the actors in this drama are dead. There are still a few of Oppenheimer's California students left, and a few of the people who were at Los Alamos with him, but nearly every week I read a new obituary. This means that I am no longer constrained by their presence, but it also means that I can no longer get their advice. I make no pretense of trying to write a "definitive" biography of Oppenheimer. I leave this to others. Rather this is *The New Yorker* profile I never wrote.

J. B.

New York City
January 2004

OPPENHEIMER

Oppenheimer with his father, Julius

1

BEGINNINGS

I ASKED I. I. RABI if he found Oppenheimer easy to get along with, and he answered, "I found him excellent. We got along very well. We were friends until his last day. I enjoyed things about him that some people disliked. It's true you carried on a charade with him. He lived a charade, and you went along with it. It was fine — matching wits and so on — and I took him for what he was. I understood his problem."

"What was his problem?" I asked.

"Identity," Rabi answered. "He reminded me of a boyhood friend about whom someone said that he couldn't make up his mind whether to be president of the B'nai B'rith or the Knights of Columbus. Perhaps he wanted to be both simultaneously. Oppenheimer wanted every experience. In that sense he never focused. My own feeling is that if he had studied the Talmud and Hebrew, rather than Sanskrit, he would have been a much greater physicist. . . ."

When Rabi first met him, Oppenheimer was twenty-five. Rabi was six years older. Both men, as it turned out, were on fellowships allowing them to study in Europe where the new quantum theory was being developed.* Both men had planned to go to Copenhagen to work with Niels Bohr. Instead both had landed in Zurich in 1929, where Wolfgang Pauli, a Viennese-born physicist, had just accepted a position. In Rabi's case, he had actually gotten to Copenhagen only to find that Bohr's institute really did not have a place for him. Without consulting him, they had arranged for him to go to Hamburg to study with Pauli. Then, after a stint in Hamburg, he followed Pauli to Zurich. In Oppenheimer's case, he had been in Leiden studying with Paul Ehrenfest, who was well known for conducting Socratic-like physics seminars. Ehrenfest felt that Oppenheimer's physics was too baroque and that going to Copenhagen would only make it worse. In an interview, Oppenheimer recalled that what influenced him was "Ehrenfest's certainty that Bohr with his largeness and vagueness was not the medicine I needed but that I needed someone who was a professional calculating physicist, and that Pauli would be right for me. . . . He thought in other words that I needed more discipline and more schooling. . . . I did see a copy of the letter he wrote Pauli . . . [and] it was clear that he was sending me there to be fixed up."

Pauli was a few years older than Oppenheimer and a few years younger than Rabi. He was one of the best theoretical physicists of his era and also a singular character. Oppenheimer once said that Pauli was the only person he knew who was identical to his carica-

*Quantum theory had been invented by people like Erwin Schrödinger, Werner Heisenberg, and Paul Dirac with the guidance of Niels Bohr. It embraced the wave and particle aspects of atomic physics, and was so successful that in the early 1930s Dirac could say that it explained most of physics and all of chemistry. But people of Rabi's generation had to go to Europe to learn it.

ture. He was a devastating critic who despised shoddy or mediocre work. Of some unfortunate physicist he said, "*So jung und schon so wenig geleistet*" (So young and already he has contributed so little). Rabi reported that he got the first shot in by saying, in reference to something Pauli had said, "*Das ist Unsinn*" (That is nonsense), which took Pauli by surprise since that was not something you were supposed to say to him. One cannot imagine Oppenheimer doing anything like this, though it is possible that Pauli was his model when it came to making nasty critical remarks, which he later did to his own students.

Oppenheimer and Rabi were in many ways polar opposites. Rabi had been born in 1898 in Rymanów in Galicia, which was a part of the Austro-Hungarian Empire. His mother tongue was Yiddish, which he said he spoke "like a native." He emigrated to the United States with his mother when he was one. His father had come a little earlier and had found work as a manual laborer. The Rabis were very poor and lived in slums, first on the Lower East Side of New York and then in Brooklyn. They were also Orthodox fundamentalist Jews. Long after Rabi had broken with this tradition he still identified himself as an Orthodox Jew. "This was the church I failed," he often noted. There was no ambiguity in how Rabi felt about anything.

Oppenheimer's father, Julius, had emigrated to the United States from Germany in 1888, when he was seventeen. His father had been a peasant farmer and grain merchant in Hanau, where Julius was born. Oppenheimer once described his grandfather as "an unsuccessful businessman, born himself in a hovel, really in an almost medieval German village, with a taste for scholarship." Julius Oppenheimer had an uncle, Sol Rothfeld, who had created a textile-importing business in New York, so he had no problem finding work in America. He also had an older brother, Emil, who had emigrated first, so he had substantial family ties here. Julius became a full partner in his uncle's firm and then started one of his own that imported lining fabrics. He had serious cultural interests, and it seems that he met his future wife, Ella Friedman, at an art exhibition. She was an artist who had studied in Europe and had taught art in New York, among other places at Hunter College. She wore a glove on her right hand that concealed a prosthetic device. She had been born with what has been described as an undeveloped or absent right hand. She was probably responsible for the superb contemporary art collection the Oppenheimers amassed. It eventually included three van Goghs: *Enclosed Field with Rising Sun*, which was still in Oppenheimer's possession when I saw it in his house in 1957; *First Steps (After Millet)*, which was acquired in 1926—it is now in the Metropolitan Museum of Art in New York; and *Portrait of Adeline Ravoux*, which is in the hands of a private Swiss collector.

Ella Friedman was also of German-Jewish background, though she had been born in New York. Her family had emigrated to the United States in the 1840s. The couple married on March 23, 1903. Julius was thirty-two and she thirty-four. They moved into an apartment at 250 West Ninety-fourth Street in Manhattan—a good address. Oppenheimer was born on April 22, 1904. His birth certificate gives his name as Julius Robert Oppenheimer. Four years later, on March 16, 1908, a second son was born—Lewis Frank Oppenheimer. He died on the first of May. His death certificate says the cause was stenosis of the pylorus. This is a congenital narrowing of part of the stomach leading to the small intestines that afflicts some babies. It is now readily treatable by surgery. The death certificate indicates there was an operation; one guesses that the baby died during its course.

One can only imagine the effect of his tiny brother's death on his parents. Oppenheimer was now even more cosseted than he had been before. In the summer of 1922, at eighteen, he made his first trip to the Southwest. In Albuquerque he met the young Paul Horgan, who was just beginning his career as a novelist. They became very close friends. Over the years Horgan recorded a number of perceptive observations of Oppenheimer and his family. After a visit to the family's summer home on Long Island, Horgan noted of Oppenheimer's parents, "She was a very delicate person . . . highly attenuated emotionally, and she always presided with a great delicacy and grace at the table and other events, but a mournful person. Mr. Oppenheimer was . . . desperately amiable, anxious to be agreeable, and I think essentially a very kind man. . . . The household was run with luxury but simplicity at the same time, every comfort and great style and charm . . . and a sadness; there was a melancholy tone." One can only imagine a child being smothered in attention; a brilliant, lonely, and precocious child, something he never really got over. In the 1930s Oppenheimer had a relationship with a woman named Jean Tatlock. She once tried to explain Oppenheimer to a friend: "You must remember that he was lecturing to learned societies when he was seven, that he never had a childhood, and so he is different from the rest of us."

There is certainly much truth in what Tatlock says, but her chronology is somewhat off. Oppenheimer recalled in an interview with the historian of science Thomas Kuhn that he had been taken to Europe by his parents to see his grandparents when he was five and again when he was seven. On one of those occasions—he did not remember which—he was given a collection of minerals. ". . . It was nothing; it was a box with maybe two dozen samples labeled in German.

"From then on I became, in a completely childish way, an ardent mineral collector and I had, by the time I was through, quite a fine collection. . . . This was at first a collector's interest, but it began to be also a bit of a scientist's interest, not in historical problems of how rocks and minerals came to be, but really a fascination with crystals, their structure, bi-refringence [if, for example, a crystal has different indices of refraction in different directions, a light beam will get split, and the split beams will propagate at different speeds], what you saw with polarized light, and all the canonical business. . . . When I was ten or twelve years old, minerals, writing poems and reading, and building with blocks still—architecture— were the other three themes that I did, not because they were something I had companionship in or because they had any relation to school but just for the hell of it. I gave up the blocks probably at the age of ten, and the minerals became a charming preoccupation. I started trying to understand them. I had very great trouble because I didn't always know the vocabulary; I think it was a month before I realized that 'intercept' could be used as a noun as well as a verb and this was the bane of me." When he was about twelve he did read a paper at the New York Mineralogical Club, in which he had become a member. This must be what Tatlock refers to. I know of no other instance.

In view of what he later became, it is curious that Oppenheimer does not mention mathematics as a childhood interest. We remember Einstein's preschool preoccupation with Euclidean geometry. The physicist Freeman Dyson once told me that when he was young enough that he was still being put down for naps, he invented for himself the notion of the convergent infinite series. He began adding up $1 + 1/2 + 1/4 + 1/16 \ldots$ and noticed that the sum was converging to 2. Hans Bethe filled notebooks with decimals and large numbers and did mental arithmetic. I have found no indication of this kind of mathematical prodigy in Oppenheimer. The stories one reads are mainly about verbal skills. Perhaps this accounts for why he was notorious for making mistakes of arithmetic factors in his formulas. By the time he was graduated from high school, Oppenheimer had studied five or six languages. I have no idea how deeply his knowledge went in any of them. To keep things in perspective, even in my day, the 1940s, in comparable New York private schools we studied three languages, including either Latin or Greek.

Julius Oppenheimer joined an entity known as the Ethical Culture Society in New York. He became a member of its board of trustees in 1907, serving under Felix Adler, its founder. It was Adler who presided over Julius and Ella's marriage. They had no temple affiliation. I doubt if Oppenheimer ever set foot in a synagogue for any religious reason. Adler had founded the Ethical Culture movement in 1876. He had come to the United States from Germany in 1857, at age six. His father, Samuel Adler, had agreed to become the rabbi of Temple Emanu-El in New York, then and now one of the most fashionable temples in the United States. It was founded by well-to-do German Jews who wanted to maintain some connection with Judaism but not Orthodoxy.

It was assumed that Felix would succeed his father. Instead he created the Ethical Culture movement. In his founding address, delivered on May 16, 1876, he said, "We propose to entirely exclude prayer and every form of ritual." Instead there would be Sunday meetings "which are to consist of a lecture mainly, and, as a pleasing and grateful auxiliary, of music to elevate the heart and give rest to the feelings." The lectures were to "contemplate large thoughts" as well as to make clear "the responsibilities which our nature as moral beings imposes on us in view of the political and social evils of our age." One of these evils was anti-Semitism, which these highly assimilated German Jews were beginning to

9

feel acutely. When asked what religion they had, they could now reply "Ethical Culture." Jews who had emigrated from Eastern Europe, such as Rabi's parents, and who had re-created a shtetl in their new country, were a source of embarrassment.

Upper-middle-class Jews like the Oppenheimers found that the better private schools would not take their children, so they created a school of their own, though it was open to anyone. The Ethical Culture School, founded in 1895, was located at 33 Central Park West. It is still there. In 1928 a second campus, Fieldston, was created in the Bronx for high school students. Oppenheimer entered the second grade in 1911 and was graduated from high school ten years later. (Rabi, of course, went to public schools.)* Several people who have written about Oppenheimer have claimed that his years at the Ethical Culture School played a great role in forming his moral and spiritual ethos. When I read this I am puzzled. Until the mid-1930s Oppenheimer had absolutely no interest in any of the ethical concerns, especially those involving public service, that were taught at Ethical Culture. In Reinhold Niebuhr's phrase, he had, to a remarkably high degree, a "neurotic preoccupation with self."† Schoolmates remembered him as a shy, physically awkward boy with not many friends. Of himself he often said, "I was an unctuous, repulsively good little boy." He was also not in very good health, which, in view of the death of his younger brother, was of enormous concern to his parents. Rabi's wife Helen attended school with Oppenheimer and recalled that by the

---

*In the 1940s my parents decided that the New York public schools might be dangerous. I was sent to Columbia Grammar, not far from Ethical Culture. A few years later Stephen Weinberg and Sheldon Glashow went to the public Bronx High School of Science. They shared the Nobel Prize for Physics.

†I heard Reinhold Niebuhr use this phrase in a sermon when I was a Harvard undergraduate. Naturally, I thought he was referring to me.

seventh grade he was universally recognized as an intellectual phe-
nomenon. But, Rabi added, "From conversations with him I have
the impression that his own regard for the school was not affectionate. Too great a dose of Ethical Culture can often sour the budding
intellectual who would prefer a more profound approach to
human relations and man's place in the universe."

Whatever Oppenheimer's feelings were about the school at
large, he found two very important teachers there; a science
teacher named Augustus Klock, and an English teacher named
Herbert Winslow Smith. Nearly a half-century later, Oppenheimer could still vividly recall the importance of these teachers.
Of Klock he told Kuhn, "I think the most important change came
in my junior year in high school. . . . The teacher of physics and
chemistry was Augustus Klock. . . . He was marvelous; I got so ex-
cited that after the first year, which was physics, I arranged to
spend the summer working with him setting up equipment for the
following year, and I would then take chemistry and would do
both. We must have spent five days a week together; once in a
while we would even go off on a mineral hunting junket as a re-
ward for this. I got interested then in electrolytes and conduction; I
didn't know anything about this but I did fiddle with a few experiments [although] I don't remember what they were. I loved chemistry so deeply that I automatically now respond when people want
to know how to interest people in science by saying 'Teach them
elementary chemistry.' Compared to physics, it starts right in the
heart of things and very soon you have that connection between
what you see and a really sweeping set of ideas which could exist
in physics but is very much less likely to be accessible. I don't
know what would have happened if Augustus Klock hadn't been
the teacher in this school, but I know that I had a great sense of in-
debtedness to him. He loved it, and he loved it in three ways: he

loved the subject, he loved the bumpy contingent nature of the way in which you actually find out about something, and he loved the excitement that he could stir in young people. In all three ways he was a remarkably good teacher."

Herbert Smith was equally important. He had been educated at Harvard where he had gotten his master's degree in English. While trying for a Ph.D. he was offered a job at Ethical Culture. He liked it so much that after his first year, 1917, the same year Oppenheimer entered the high school, he stayed. All throughout his Harvard years—one would imagine it was Smith who encouraged his choice of university—Oppenheimer wrote letters to Smith. Interestingly, they are signed "Bob"—at least in the beginning. After he had been at Harvard for a year or so he began signing "Robert" or "R." In later years, letters to good friends would be signed "Oppie" or "Opje," a nickname he had picked up in Leiden. More formal letters were signed "J. Robert Oppenheimer." I have never found a letter or any other document that he signed "Julius." He often said it was "'J' for nothing"—an odd way to refer to his first name, which was the name of his father. Equally odd is that in all the FBI reports signed by J. Edgar Hoover, Oppenheimer is referred to as "Julius Robert Oppenheimer."* The FBI must have taken the trouble to look at his birth certificate.†

*Hoover's first name was John.

†Curiously, the authors I have read on Oppenheimer's life have not done this. You will find "J for nothing" in Herken, Stern, Goodchild, Royal, and Michelmore. Smith claims it to be an unresolved mystery and cites Oppenheimer's brother Frank, who is in the "J for nothing" school. Particularly amusing to me is a footnote in Stern which reads, "There is some disagreement about the 'J' in Oppenheimer's name. Official communications by the FBI director J. Edgar Hoover, as well as the 1968 book *Lawrence and Oppenheimer*, by Nuel Pharr Davis, referred to him as 'Julius Robert Oppenheimer.' In a review of the Davis book, Jeremy Bernstein, a close friend of Oppenheimer, objected that 'the "J" doesn't

On August 14, 1912, Oppenheimer's brother Frank Friedman Oppenheimer was born. His mother was in her early forties. Before Frank Oppenheimer's birth the family had moved from Ninety-fourth Street to a large apartment on the eleventh floor of 155 Riverside Drive, an even better address. The apartment was large enough so that two servants lived in as well as Ella Friedman's mother, Cecilie. I think the arrival of a younger brother was as important to Oppenheimer's development as were his teachers. If one reads his letters from, say, his Harvard years, they divide into two groups—those to Frank and those to everyone else. The letters he wrote to Smith, for example—and we will see an example—are extremely mannered. Perhaps what he is describing happened, but certainly not in the way he described it. He is inventing a persona that he thinks his teacher will admire. The letters to Frank, on the other hand, are filled with affection and good sense. There is no pretense at all. One must remember that as Frank moved into his teens, his parents were already deep in middle age. Oppenheimer was almost a father as well as a brother. This was important because it is quite clear that, even when his parents were younger, communication with them was difficult. When he was fourteen, Oppenheimer went to a summer camp in New York State for well-

stand for anything.' This conforms with what Oppenheimer himself is known to have told associates. However, the Bureau of Records and Statistics, Department of Health, City of New York advises that the city's records 'contain a birth certificate for a Julius R. Oppenhiemer' [sic] born on April 22, 1904." I made this rash statement before I had actually seen the birth certificate. I do not find, incidentally, the misspelling of Oppenheimer's name in the original. The books referred to here are Gregg Herken, *Brotherhood of the Bomb* (New York, 2002); Philip M. Stern, *The Robert Oppenheimer Case* (New York, 1969); Peter Goodchild, *J. Robert Oppenheimer: Shatterer of Worlds* (Boston, 1981); Denise Royal, *The Story of J. Robert Oppenheimer* (New York, 1969); Peter Michelmore, *The Swift Years: The Robert Oppenheimer Story* (New York, 1969); and Nuel Pharr Davis, *Lawrence and Oppenheimer* (New York, 1986).

to-do Jewish boys. It was run by a man named Otto Koenig, who was the principal of Sachs Collegiate Institute in New York. The experience, to put it mildly, was not a success. At one point he wrote back to his family that he was learning the facts of life. His parents complained, and the camp officials tried to censor the telling of dirty jokes—a mainstay in a camp like this.* This outraged his fellow campers, who locked him in an icehouse without his clothes after having painted his bottom green. I have never read an interview with Oppenheimer in which he mentions this. It must have been an awful experience.

Oppenheimer was admitted to Harvard in 1921. This, as it happened, was a year before A. Lawrence Lowell, the president of the university, tried to impose a quota on the number of Jewish students. This attempt inspired so much adverse publicity that the university replaced a Jewish quota with a geographical quota, which accomplished the same thing—to limit the number of Jewish undergraduates to no more than 15 percent. Lowell also had the notion that the faculty should consist of urbane scholars of the British type, a formula that did not fit many Jews. One wonders whether Oppenheimer would have been admitted to Harvard if he had applied a few years later. As it happened, he was unable to begin in 1921. That summer his parents had taken him to Europe, and he had acquired a form of dysentery that nearly killed him. He spent the next year recovering. During the summer of 1922, in fact, his father hired Herbert Smith to accompany Oppenheimer on a recuperation trip to the Southwest. Oppenheimer asked Smith to introduce him as his brother while on the trip, which Smith interpreted as an attempt by Oppenheimer to conceal his Jewishness.

*I speak from experience, having spent several summers in such a camp connected to my school.

Smith refused. It was on this trip, incidentally, that Oppenheimer learned to ride horses. One of the rides he made was to the Los Alamos mesa.

In the fall of 1922, Oppenheimer entered Harvard. This seems to have been a transforming experience—an intellectual banquet. He wrote often to Smith, sending along poems or short stories. Here is an excerpt from a letter written in Oppenheimer's second year. (One must be careful in his case about using terms like "sophomore" or "junior," since Oppenheimer graduated in three years.) He writes,

". . . I begin to believe in eternal passions now, when I see that each note from you still sends me into a violent schoolgirl flutter of excitement. [I don't think anything should be read into this language. It is Oppenheimer exuberance.] I am able, though, to assure myself that my delight at your last one was at least half due to your magnanimous notes on my verses. I think you are very right about them, though perhaps a trifle too kind. I detect in your 'you write for the thrill' the old rebuke of handshake allegory; but I think the real crime lies, not there, but in the shameless exhibitionism which lets me reveal the debauches. Of course, when I do scrawl, I am always able to justify my strained figures and racked words; and not infrequently by a puny etymological device: limbus—limbo (that's true)—borderland of hell; or even worse, palindrome—running both ways—repetition. You see? It's vile, I admit. Generously, you ask what I do. Aside from the activities exposed in last week's disgusting note, I labor, and write innumerable theses, notes, poems, stories, and junk; I go to the math lib and read and to the Phil lib and divide my time between Meinherr [Bertrand] Russell and the contemplation of a most beautiful and lovely lady who is writing a thesis on Spinoza—charmingly ironic, at that, don't you think? I make stenches in three different labs, listen to

Allard [Louis Allard was a professor of French] gossip about Racine, serve tea and talk learnedly to a few lost souls, go off for the weekend to distill the low-grade energy into laughter and exhaustion, read Greek, commit faux pas, search my desk for letters and wish I were dead. Voila. . . ."

Smith must have seen something special in Oppenheimer to have kept a letter like this. The whole tone makes one's flesh creep. Furthermore it is a fantasy. The reality is that Oppenheimer was a socially isolated scholar who was taking six courses a year to graduate in three. One wonders if he ever had a date during this time. "I wish I were dead" probably comes closest to his real feelings. Perhaps Smith understood this. At the risk of getting a bit ahead of the chronology, I quote from a 1926 letter Oppenheimer wrote to his brother, who had now also entered Ethical Culture. In the way of younger brothers, Frank was trying to emulate Robert in his choice of studies. Oppenheimer writes,

". . . Mr. Klock told me that he was enjoying his work with you.

"And still more advice: I don't think you would enjoy reading about relativity very much until you have studied a little geometry, a little mechanics, a little electrodynamics. But if you want to try, Eddington's book is the best to start on. [Arthur Eddington was a noted British astronomer who wrote a classic text on relativity. When asked if it was true that only three people in the world understood the theory, he responded, "Who is the third?"] I remember five years ago you were dressed up to act like Albert Einstein; in a few years, it seems, they won't need to disguise you. And you will be able to write your own speech. And now a final word of advice: try to understand really, to your own satisfaction, thoroughly and honestly, the few things in which you are most interested; because it is only when you have learnt to do that, when you realize how hard and how very satisfying it is, that you will appreciate fully the more spectacular things like relativity and mechanistic biology. If you think I am wrong please don't hesitate to tell me so. I'm only talking from my own very small experience. . . ."

That is a letter worth keeping.

In June 1925, Oppenheimer was graduated summa cum laude, Phi Beta Kappa. His major was chemistry, though by this time he was really interested in physics. Here again he had two influential teachers, Percy Bridgman and Edwin Kemble. Bridgman, who was born in Cambridge, Massachusetts, in 1882 (he died in 1961), was himself a Harvard graduate. Throughout his long career he had two interests, the physics of high pressures and the philosophy of science. He won the 1946 Nobel Prize for Physics for his research. In 1927 he published a highly influential book called *The Logic of Modern Physics* in which he introduced a moderate form of positivism that came to be called "operationalism." The idea was that concepts in physics such as mass and charge could only be defined in terms of the operations needed to measure them. If one could not do that, it meant that the concept was meaningless. ("Time," for example, meant nothing more or less than the operations performed on or by a clock.) Bridgman took on very few students, and Oppenheimer became a kind of tutee. He tried unsuccessfully to become an experimental physicist.

Kemble, who was born in 1889 and died in 1984, came to Harvard in 1913 as a graduate student. He too worked with Bridgman, who became his thesis adviser. Because Harvard, like most American universities at that time, did not really have theoretical physics, which is what Kemble wanted to do, he was pretty much on his own. Indeed, the department insisted that his thesis have an experimental component. While Bridgman did not know quantum theory, he understood its importance and persuaded Kemble to come back to Harvard in 1919 to try to build up theoretical physics. Kemble was teaching most of the theory courses when Oppenheimer was an undergraduate. By his own admission he was not a top-notch physicist, but he was a top-notch teacher. His first graduate student, John van Vleck, later won the Nobel Prize.

As an aside, I encountered both Kemble and Bridgman when I was at Harvard. I had written—and indeed published—a short article in a philosophy journal in which I took issue with something involving operationalism. I have no recollection what. In any event, Bridgman asked to see me. I went down to his lab in the basement of Jefferson Hall and found him wearing some kind of smock. He told me that I had been kinder to him than many of his critics. We then had a most amiable discussion. I was a sophomore. My association with Kemble was more long-standing. I was a teaching assistant in a course he taught with the historian of science Gerald Holton, also a Bridgman student. It was designed for undergraduates who were not majoring in the sciences. I had several discussions with Kemble. In one of them he told me that if I wanted to lead a happy life I should make sure that my ambitions and abilities were comparable. I wonder if he told this to Oppenheimer.

At this time any American physicist who wanted to be where things were really happening in physics had to go to Europe. After consulting with Bridgman, Oppenheimer decided he wanted to go to the Cavendish Laboratory at Cambridge University to get his Ph.D. The master of the Cavendish was Ernest Rutherford, one of the greatest experimental physicists who ever lived. He and his associates had discovered the atomic nucleus and had been engaged in elucidating its structure. Bridgman wrote a letter of recommendation to Rutherford. The first part is often quoted. After commenting on Oppenheimer's evident brilliance, Bridgman went on to say,

"His problems have in many cases shown a high degree of originality in treatment and much mathematical power. His weakness is on the experimental side. His type of mind is analytical, rather than physical, and he is not at home in the manipulations of the laboratory. . . . During his last year he started with me a small research on the effect of pressure on the resistance of alloys, and was evidently much handicapped by his lack of familiarity with ordinary physical manipulations. However, he stuck at it, and by the end of the year had learned much, and obtained some results of value, all without being of much trouble to me personally, but he picked up many of the tricks of manipulation which he needed from my mechanic.

"It appears to me that it is a bit of a gamble as to whether Oppenheimer will ever make any real contributions of an important character, but if he does make good at all, I believe that he will be a very unusual success, and if you are in a position to take a small gamble without too much trouble, I think you will seldom find a more interesting betting proposition."

But then there was a final paragraph which is less often quoted. Bridgman wrote, "As appears from his name, Oppenheimer is a Jew, but entirely without the usual qualifications of his race. He is a tall, well set-up young man, with a rather engaging diffidence of manner, and I think you need have no hesitation whatever for any reason of this sort in considering his application."

It must be understood that Bridgman was not an anti-Semite. If he had been, he would not have taken on Oppenheimer as his student. He was trying to be helpful, and he was reflecting the social climate of the day. As it happened, the letter did not succeed. The reason Rutherford gave was that he took on only people who came with a well-defined research agenda. This seems unlikely since Rutherford's "boys" were discovering things for which there was no agenda. Possibly Rutherford thought of Bridgman as someone working in a backwater. How can one compare discoveries in high-pressure physics with the discovery of the atomic nucleus and its properties?

But Oppenheimer was not easily discouraged. He sailed for England in September 1925, having been admitted to Christ's College in Cambridge. He managed to find a working space in the Cavendish basement and some supervision by the experimental physicist J. J. Thomson, who was nearby. By this time Thomson, who had been born in 1856, was, in physics terms, an antique. His discovery of the electron had happened before the turn of the century. He had retired from his professorship and was now an honorary professor, which presumably gave him the right to some space in the laboratory. As a guide to the quantum mechanics revolution that was about to happen, he must have been less than useless. This conspired with a host of other things—squalid living quarters, the absence of a nurturing family and a familiar environment—to produce in Oppenheimer an acute nervous breakdown. Whatever dam had been holding his psyche together burst.

The most dramatic manifestation of this occurred on a vacation trip to Paris in the fall of 1925. There Oppenheimer met with a good friend, Francis Fergusson. In the course of a conversation about personal matters he suddenly made an attempt to strangle Fergusson. This was not a joke. Fergusson was able to fight him off. This episode has never really been explained, though Oppenheimer referred to it somewhat jocularly in a letter to Fergusson in March 1926: "My regret at not having strangled you is now intellectual rather than emotional," he wrote. There is some suggestion that it might have had to do with feelings of sexual inadequacy. Oppenheimer did see a couple of psychiatrists in England but claimed they were useless. What seems to have jolted him out of this state was a hiking trip to Corsica with a couple of friends. The trip was to continue to neighboring Sardinia when Oppenheimer suddenly announced that he had to return to Cambridge because he had left "a poisoned apple" on the desk of P. M. S. Blackett, an experimental physicist who was some years Oppenheimer's senior. His friends had no idea what to make of this. I find it very unlikely that it was anything more than the mythmaking Oppenheimer indulged in for most of his life, sometimes with disastrous consequences for himself and others. He must have scarcely known Blackett, who had just arrived back in Cambridge after a sojourn in Göttingen. In any event, Oppenheimer did return to England.

Despite appearances, Oppenheimer always had both physical and mental resilience. In this instance, despite everything, he was able to produce two serious papers on the applications of the just discovered quantum theory—his first scientific publications. Among the people these papers impressed was Max Born, who was visiting Cambridge. Born, who died in 1970, was one of the lesser-known heroes of the discovery of quantum mechanics.* It took until 1954 before he was awarded the Nobel Prize for Physics for work he had done in the 1920s. Werner Heisenberg, Erwin Schrödinger, and Paul Dirac are the more recognizable heroes. When Heisenberg discovered the so-called "matrix mechanics" in 1925, he did not know what a matrix was. He had found some expressions that appeared to combine in a curious algebraic fashion. Born explained to him that what he was doing was matrix algebra. Quantities like position and momentum are represented by these matrices. This ultimately leads to the Heisenberg uncertainty principle—limitations on the measurement of these quantities. When

*An endearing fact about Born is that he was the grandfather of the Australian popular singer Olivia Newton-John.

Schrödinger discovered, in 1926, the equation for the waves that bear his name, he had a misconception as to what these waves represented. He thought they were material waves that guided the motion of particles. It was Born who explained that they were waves of probability. What they expressed was the probability of locating particles. It was to Born that the somewhat older Einstein wrote to say that he did not believe that God played dice. Beginning in 1921, Born created a school of theoretical physics in Göttingen that attracted several of the most brilliant young physicists of the day. I suspect it was the model on which Oppenheimer built his own school in California in the 1930s. In any event, Born invited Oppenheimer to come to Göttingen to complete his work on his Ph.D. It turned out to be a transforming experience.

Much of this transformation had to do with working with Born, the first real theorist Oppenheimer ever encountered at close range. But he also met people his own age who were at least as smart and, in some cases, vastly more creative in physics than he was. Foremost among these was Paul Dirac. Dirac was just two years older, but in his Ph.D. thesis he had invented the version of the quantum theory that even now one would teach to somewhat more advanced students. If Pauli was identical to his caricature, Dirac was beyond caricature. The original was too good to be true. At the mention of his name, one is always tempted to stop the narrative and recount Dirac stories. I will limit myself to two, both of which are related to Oppenheimer. The first one I was witness to.

In the late 1950s when I was at the Institute for Advanced Study Dirac was also a visitor. He spent much of his time in the woods near the Institute with an axe, chopping a trail in the general direction of Trenton. But he had an office in the building in which a number of us also had offices. Oppenheimer had decided that temporary members like myself would be better off without a telephone in our offices; telephone calls might interrupt the contemplation we were supposed to be engaged in. Thus there was a telephone in the hall, and when it rang the subsequent conversation interrupted the contemplation of everyone on the floor. One of these phone calls was for Dirac. As it happened, the late Abraham Pais, a permanent member of the Institute, was in my office discussing some point of physics. We were interrupted by Dirac's phone conversation which consisted largely of yes's and no's and lapses into silence. Dirac was notorious for his economy of words. When the call was over, Dirac appeared in my office. He ignored me completely and spoke directly to Pais. He explained that the phone call was from a reporter for the *New York Times* who wanted an advance copy of a talk that Dirac was to give in New York later in the week. Dirac's concern was that the talk might be published before he gave it. Pais suggested sending it with a written caveat saying, "Do not publish in any form." Dirac then stood there evidently deep in thought. After several minutes Pais and I went on with our discussion while Dirac still stood thinking. With no warning Dirac suddenly spoke. "Isn't 'in any form' redundant in that sentence?" he asked.

The second story is one that Oppenheimer often told on himself, always with great affection. During his stay in Göttingen he and Dirac lived in the same pensione. They often went for evening walks. On one of these walks, which was on the walls that surrounded the old university town, Dirac began to chide Oppenheimer for the time he spent trying to write poetry. "How can you do both poetry and physics?" he asked. "In physics we try to tell people things in such a way that they understand something that nobody knew before. Whereas in poetry . . ." Dirac, like so many other people who knew Oppenheimer, understood that he lacked the single-mindedness—not the ability—to do really great physics.

Dirac was, if anything, single-minded. Once when he was to go on a long sea voyage, someone suggested a book he might like to read. He remarked that he didn't read books because they interfered with thought.

In Göttingen, Oppenheimer had another of those failed romantic experiences that seemed to be part of his life. Apparently he met a fellow student named Charlotte Riefenstahl on a train when both of them were traveling from Hamburg to Göttingen with a group of students. All of them had the usual grotty luggage except for one immaculate pigskin suitcase. Riefenstahl inquired and was informed that it belonged to Oppenheimer. She was also told not to admire it because then he would give it to her. It seems that when someone admired something he had, Oppenheimer felt obliged to give it away. Nonetheless, when Riefenstahl was leaving Göttingen, Oppenheimer appeared at her flat with the suitcase. He also assiduously tried to court her—too assiduously. In 1931 she married another fellow student, Friedrich Houtermans. Pauli was one of the witnesses at their wedding. Houtermans later had the unfortunate distinction of being arrested in Russia by the NKVD and then being sent back to Germany where he was promptly arrested by the Gestapo. He was released and went to work for an entity devoted to nuclear energy. One of the things he did was to suggest the use of plutonium for a nuclear fuel. As for the pigskin bag, it seems as if Riefenstahl used it until it wore out.

The problem that Born and Oppenheimer worked on—which became Oppenheimer's Ph.D. thesis—had to do with the treatment of molecules in quantum mechanics. The treatment of simple atoms like hydrogen was pretty much under control, but molecules, which might have several nuclei and their attendant electrons, were a much more difficult problem. It was the quantum mechanical equivalent of computing the orbits of a system involving several bodies that interact with one another. It was essential to treat because most of the objects that enter into chemical reactions are molecules, not single atoms—water being a familiar example. The approximation that Born and Oppenheimer introduced is with us still and is a standard part of any quantum mechanics course. They noted that in a molecule the atomic nuclei are so much more massive than the electrons that they can be considered as stationary spectators. They essentially drop out of the problem. Then one can treat the electrons separately—difficult but not impossible. This has become the standard technique for dealing with molecules. On the basis of this work Oppenheimer received his Ph.D. from the University of Göttingen in March 1927. For many years people who wrote about his time with Born wrote as if their relationship with him was an even closer version of Oppenheimer's relationship with Bridgman. But then in 1975, Born's posthumous autobiography, *My Life*, was published. Its candor came as something of a surprise to many people. By this time Oppenheimer was dead. This is what Born wrote:

"Oppenheimer caused me greater difficulty [Born had just described his relations with some of the other students]. He was a man of great talent and he was conscious of his superiority in a way which was embarrassing and led to trouble. In my ordinary seminar on quantum mechanics, he used to interrupt the speaker, whoever it was, not excluding myself, and to step to the blackboard, taking the chalk and declaring: 'This can be done much better in the following manner . . .' I felt the other members did not like these personal interruptions and corrections. After a while they complained, but I was a little afraid of Oppenheimer, and my halfhearted attempts to stop him were unsuccessful. At last I received a written appeal. I think that Maria Göppert, then a very young student, now a well-known professor in California, was the driving force. [Maria Göppert Mayer, who died in 1972, had gone from the University of Chicago to La Jolla in 1960. In 1963 she won the Nobel Prize for Physics for her theoretical work on the structure of nuclei. Since Born knew about her move to California but did not mention the Nobel Prize, I would guess that these observations must have been written sometime between these two events.] They gave me a sheet of paper looking like parchment in the style of a medieval document, containing the threat that they would boycott the seminar unless the interruptions were stopped. I did not know what to do. At last I decided to put the document on my desk in such a way that Oppenheimer could not help seeing it when he came to discuss the progress of his thesis with me. To make this more certain, I arranged to be out of the room for a few minutes. This plot worked. When I returned I found him rather pale and not so voluble as usual. But I am afraid he was deeply offended, though he never showed it and he gave me a splendid present before he left, a first edition of Lagrange's *Mécanique Analytique*, which I still possess. But I was never invited to the U.S.A., in particular not to Princeton after the end of the Second World War, when every physicist of any standing spent some time at the Institute for Advanced Study, whose president [director] Oppenheimer had become. But this is only a conjecture, and it is quite possible that neither his influence nor his rancour was as great as I imagined. Maybe the reason for my being neglected was the well-known fact that I was opposed to atomic weapons and criticized those who made them."

My guess would be that by this time Born's physics was very far out of the mainstream. So was Dirac's. But Dirac was Dirac.

For the next two years Oppenheimer visited various institutions on fellowships. He spent the fall semester of 1927–1928 at Harvard.

He was still writing poetry, and in June 1928 he published the only poem of his that I have ever seen in print. The previous fall two Harvard freshmen, Varian Fry and Lincoln Kirstein, had founded a literary magazine they called *Hound and Horn*. It lasted seven years and became quite influential. Oppenheimer's poem was called "Crossing."

It was evening when we came to the river
with a low moon over the desert
that we had lost in the mountains, forgotten,
what with the cold and the sweating
and the ranges barring the sky.
And when we found it again,
in the dry hills down by the river,
half withered, we had hot winds against us.
There were two palms by the landing;
the yuccas were flowering; there was
a light on the far shore, and tamarisks.
We waited a long time, in silence.
Then we heard the oars creaking
and afterwards, I remember,
the boatman called to us.
We did not look back at the mountains.

By this time Oppenheimer could have had his pick of universities both in the United States and in Europe. Kemble visited Göttingen while Oppenheimer was there and reported back to Harvard that Oppenheimer had turned out to be even more brilliant than anyone had thought. He certainly could have gotten a position at Harvard. But he took a combined appointment at the California Institute of Technology and Berkeley. Cal Tech had a reasonably well-developed theoretical physics program while Berkeley didn't. It was a *tabula rasa*. This is what appealed to Oppenheimer. He could create a school.

Rabi told me that when he and Oppenheimer were together in Germany they often talked about the disrespect—the contempt—for American physics. Rabi said that in Hamburg so little was thought of the journal of the American Physical Society, *The Physical Review*, that the librarian uncrated the issues only once a year. They decided that something had to be done about this. Each built a great American school of physics—Rabi at Columbia and Oppenheimer at Berkeley. When the Second World War came, they and their students were ready to help man the laboratories that created radar and built the atomic bomb.

In the fall of 1929, Oppenheimer went to Berkeley. That summer he and his brother were able to rent a ranch in the upper Pecos Valley in New Mexico, not far from Santa Fe. Eventually they bought it. It now belongs to Oppenheimer's son Peter. When Oppenheimer first heard that the ranch was for rent he shouted "Hot dog!" The woman from whom they were renting it said that, because of their Spanish-speaking neighbors, it should be "Perro Caliente," and that became its name.

Oppenheimer with Ernest Lawrence at Perro Caliente

CALIFORNIA DAYS

PROF. TAKES GIRL FOR RIDE; WALKS HOME

J. Robert Oppenheimer, 30, associate professor of physics at the University of California, took Miss Melva [sic] Phillips, research assistant in physics living at 2730 Webster Street, for an automobile ride in the Berkeley Hills at 3 o'clock this morning.

He stopped his machine on Spruce Street at Alta Street and tucked a large robe about his passenger.

"Are you comfortable?" Prof. Oppenheimer asked.

Miss Phillips replied that she was.

"Mind if I get out and walk for a few minutes?" he queried.

Miss Phillips didn't mind, so the professor climbed from the auto and started to walk.

One hour and 45 minutes later Patrolman C. T. Nevins found the professor's car and Miss Phillips still comfortable, dozing in the front seat. He woke her and asked for an explanation of her early morning nap.

Miss Phillips told her story. Police headquarters was notified that Prof. Oppenheimer was missing and a search was launched.

A short time later the professor was awakened from a sound sleep in his room at the Faculty Club, two miles distant from his auto, and asked to explain.

"I am eccentric," he said.

— Berkeley newspaper, 1934

AFTER TAKING her undergraduate degree from Oakland City College in Indiana in 1926, Melba Phillips, who had been born in southern Indiana to a family of farmers and schoolteachers, taught high school for a year. She then went to Battle Creek College in Michigan in 1927, where she studied advanced mathematics and physics. In the summer of 1929 she went to a summer school in Michigan where she first learned quantum mechanics. One of the teachers there, E. U. Condon, a well-known American theoretical physicist, suggested that she go to Berkeley to do her doctorate. She did not go there to study with Oppenheimer especially, she told me, but he was the professor who was teaching the quantum theory in which she wanted to work. "He was very bright," she added.

Oppenheimer had three other students when Melba Phillips arrived. She was the only woman working toward a degree in theoretical physics, but there were several doing experimental physics. She took her degree in 1933. She was able to stay on for two more years by doing a variety of part-time jobs such as grading papers. It was during this time that she and Oppenheimer collaborated on a paper—"Note on the Transmutation Function for Deuterons"— which introduced what became known as the Oppenheimer-Phillips process. Remarkably, nearly three-quarters of a century later, there are several websites devoted to it. It is a nice bit of nuclear physics, and it also demonstrates Oppenheimer's ability to work with experimenters, something that was to be invaluable at Los Alamos. Not all theorists can, or want, to do this, but those who can do it well are worth their weight in gold, especially if they can give clear theoretical explanations of experimental results. Oppenheimer functioned during this period as the resident theorist for the Berkeley experimenters, who included Ernest Lawrence and Edwin McMillan. Lawrence, as we shall see, came to play a profound role in Oppenheimer's life.

In 1930, Lawrence invented the cyclotron. This device accelerates charged particles in circles that are maintained by a magnetic field. It enables these particles to reach speeds that, in more powerful cyclotrons, begin to approach that of light. McMillan, who had received his degree from Princeton in 1932, was persuaded by Lawrence to come to Berkeley to work on the cyclotron.

In 1931, not long before McMillan moved to California, Harold Urey at Columbia University managed to isolate a new form of hydrogen. Ordinary hydrogen has as its nucleus a single, positively charged particle, the proton. Heavy hydrogen, which is what Urey was interested in, has a proton and a neutral partner—the neutron—in its nucleus. This nucleus is called the deuteron. A still heavier isotope has two neutrons and a proton in its nucleus, which is called a triton. Urey detected "heavy water," a molecule in which two deuterons are chemically bonded to an oxygen atom, which is present in natural water in spectroscopically detectable amounts. Once the deuteron had been discovered, it was natural to try to make a deuteron beam to run in Lawrence's cyclotron, and in 1934 McMillan achieved it. By present standards it was a fairly primitive low-energy beam, but it represented a real breakthrough. An entirely new experimental technique for doing nuclear physics became available.

McMillan and his collaborators wanted to irradiate various elements with the beam and so produce new radioactive isotopes. They had a very clear idea of how this should come about. This involved an important principle of quantum mechanics: the wave nature of matter. A nucleus consists, in addition to its allotment of electrically neutral neutrons (one in the case of the deuteron), of a number of protons—the positively charged counterpart to the neutron. The deuteron, as noted, has one proton and one neutron. In general, protons give nuclei their positive electric charge. But the deuteron itself has a positive charge. We know that like charges repel, so if this was all there was to it, the deuteron would not be able to penetrate the target nucleus. It would bounce off like a small stone being thrown—without too much force—against a windowpane. But the deuteron has a wave nature which it exhibits under suitable experimental conditions. This changes everything. A light wave, for example, coming up against a window will be transmitted—at least partially—through the glass. This is what happens to the deuteron when it comes up against the barrier created by the positive electric charges of the nucleus. In quantum mechanics this is known as barrier penetration, or quantum mechanical tunneling. The particle "tunnels" through the barrier. It was discovered in the late 1920s. In 1928, in fact, Oppenheimer had written the first paper about it.* He then applied the idea to the extraction of electrons from metal surfaces. The electrons are held to the surface by a strong electric force. In the absence of the tunneling effect it would take a large electric field to extract them. Be-

---

*This does not seem to be widely understood. The credit is usually given to George Gamow, Ronald Gurney, and Edward Condon, who applied tunneling to the decay of some unstable nuclei by the emission of so-called alpha particles, which are in fact nuclei of helium. This is the example we use when we teach the subject. But Oppenheimer's paper was submitted several months earlier.

cause of tunneling, a weak field works. This is the principle of the scanning tunneling microscope which was developed in 1982.

The tunneling theory makes a clear prediction as to how rapidly the probability of transmission through the barrier increases as you raise the energy of the incoming deuteron. To test this, and to create new isotopes, in 1935 Lawrence, McMillan, and a physicist named R. L. Thornton did the experiment. Much to their surprise, the increase with energy was less than the theory predicted. This was the dilemma that Oppenheimer and Phillips were presented with. They realized that, as nuclei go, the deuteron is large and loosely bound. In fact, when it impinges on a target nucleus its proton can be considered as a spectator while its neutron gets stripped away since it has no trouble penetrating the charge barrier. It tunnels right into the nucleus, creating a new isotope, which may have interesting or useful properties.

In the spring of 1935, Oppenheimer, who was then in Pasadena on his annual pilgrimage to Cal Tech, was able to write to Lawrence, "I am sending Melba today an outline of the calculations & plots I have made for the deuteron transformation function. The analysis turned out pretty complicated & I have spent most of the nights this week with slide rule and graph paper." The idea of Oppenheimer spending his nights doing with a slide rule something that can now be done in a few minutes with any PC is quite moving. Phillips was given the task of checking and generalizing his calculations. The process they proposed has remained a valuable tool for probing the structure of nuclei. (During a summer in which I was an intern at the Harvard cyclotron in the early 1950s, I worked on a version of this same experiment. I have an ineluctable recollection of literally sewing together with needle and thread the targets we were going to use. At the time I did not realize who had done the first calculations of the theory.)

By the time Melba Phillips left Berkeley for a distinguished career, largely in physics education, Oppenheimer had established what became a great school for theoretical physics. He had also created a legend, intentionally or unintentionally, around his personality. There are many testimonials to this, some by himself. On the first day of his hearing before the Personnel Security Board of the Atomic Energy Commission in April 1954—the hearing in which he lost his security clearance to advise the government on problems connected with nuclear energy—Oppenheimer presented what amounted to a sort of autobiography. At one point he told the committee,

"My friends, both in Pasadena and Berkeley, were mostly faculty people, scientists, classicists, and artists. I studied and read Sanskrit with Arthur Rider [sic, Arthur Ryder, a professor of Sanskrit at Berkeley who made a number of translations of Sanskrit classics]. I read very widely, must [sic, perhaps "just"—there are a number of obvious errors in these transcripts] mostly classics, novels, plays, and poetry; and I read something of other parts of science. I was not interested in, and did not read about economics or politics. I was almost divorced from the contemporary scene in this country. I never read a newspaper or a current magazine like Time or Harper's; I had no radio, no telephone; I learned of the stock market crash in the fall of 1929 only long after the event; the first time I even voted was in the presidential election of 1936. To many of my friends my indifference to contemporary affairs seemed bizarre, and they often chided me with being too much of a highbrow. I was interested in man and his experience. I was deeply interested in my science; but I had no understanding of the relations of man to his society."

Reading this, one might get the impression that Oppenheimer was living in a kind of monastic order. But much has been left out.

In the first place, he was quite well to do. He had an academic income of about $5,000 a year and a private income of another $10,000.* To put this in perspective, Einstein, who must have had the highest income of anyone in academic physics, was offered $15,000 a year as a professor at the Institute for Advanced Study. At the other end of the scale, Robert Serber, who came to Berkeley in 1934 as a postdoctoral and was to become one of Oppenheimer's closest associates, had won one of five National Research Council Fellowships in theoretical physics. It paid $1,200 a year, and Serber was married. Oppenheimer was a bachelor. His parents were wealthy, so he had no one to support but himself.

Not long after his outing with Melba Phillips, Oppenheimer moved from the faculty club to an apartment on Shasta Road in Berkeley. He often entertained with social evenings that began with mint juleps and ended in an expensive restaurant. If the evening included students or postdocs, Oppenheimer paid the bill. In his delightful memoir *Peace and War*, Serber describes the time when the Berkeley group was joined by the Swiss-born theoretical physicist Felix Bloch, who was at Stanford and would later win the Nobel Prize. Bloch, after a dinner at Jack's, Oppenheimer's favorite restaurant in San Francisco, following a joint seminar, decided that he would like to pick up the check. He took one look and put it back.

Unlike his students, Oppenheimer had a taste for good wines and the money to pay for them. He also had a vast array of general culture in several languages that his students did not share. In the beginning this crept into his physics lectures which, while elegant, the students found hard to follow. Serber once told me a story that

---

*This figure refers to an income tax form in the early 1940s, so at this time his income may have been somewhat less. Nonetheless, by the standards of the day it was substantial.

seems typical. When Oppenheimer went to Pasadena he asked one of his graduate students to lecture for him at Berkeley, noting that the material could be found in a book. It turned out that the book was in Dutch, and when the student complained Oppenheimer remarked, "But it's such easy Dutch." In the course of a few years he became a better—even a brilliant—lecturer. Students took his quantum mechanics course several times. Serber reports on a Russian woman who took it three times and wanted to take it a fourth. When Oppenheimer refused, she went on a hunger strike until she was allowed back into the course.

Unlike, say, Enrico Fermi's University of Chicago lectures, as far as I know Oppenheimer's notes were never published. The closest one comes is a quantum mechanics text by the theoretical physicist Leonard Schiff. In the preface he expresses "his appreciation to Prof. J. R. Oppenheimer for introducing him to several of the ideas and examples which helped to give the book its form." It is the book from which many of the physicists of my generation learned quantum mechanics, so we are all students of Oppenheimer.

Serber has described Oppenheimer's way of working with students. Typically he had eight or ten graduate students and a half dozen postdoctorals—a very large number for a single professor to handle. Every day they would meet collectively with Oppenheimer, who would review the progress of each of them and make suggestions. (When he became director of the Institute for Advanced Study after the war, he used to say that its motto was, What we don't understand, we explain to each other.) When Serber became Oppenheimer's assistant, his job was to explain to the various participants in these sessions what Oppenheimer had just said. Most of them seem to have thrived on this method, but a few did not. One of the exceptions was Julian Schwinger, who became one of the towering figures in twentieth-century theoretical physics. Schwinger was a prodigy. When he entered the City College of New York in 1934, he wrote his first physics paper while still a freshman. He had taught himself most of the curriculum in physics and mathematics while still in high school. The problem was that he was failing most of his courses. He had no time for things like writing papers for his English class. It is unclear what would have happened to him if he had not been rescued by Rabi. A friend of Schwinger's at City College had taken him for a visit to Columbia and had gotten into a physics argument with Rabi. Schwinger suddenly piped up and settled the question. Rabi thought, "Who is this kid?" and proceeded to get him admitted to Columbia with a scholarship. Under Rabi's firm hand, Schwinger blossomed and was even graduated Phi Beta Kappa. He then stayed on at Columbia to get his Ph.D. and, after a year at Wisconsin, came back to Columbia to work with Rabi. In 1939, Schwinger won a National Research Council fellowship and chose to go to Berkeley to join Oppenheimer's group.

Oppenheimer was not prepared for Schwinger's work habits. Schwinger worked all night and slept all day. He continued to do this for the rest of his career. After the war, when he was at Harvard and was doing the calculations in quantum electrodynamics that led to his 1965 Nobel Prize, he carelessly dropped a lighted cigarette in a wastebasket filled with papers. He never noticed the developing fire, which was put out by the night cleaning lady. Schwinger was hardly one who would go to a group meeting of the kind that Oppenheimer organized, especially if it was held during the day. At first Oppenheimer was disappointed with Schwinger and thought of asking him to leave. But he soon realized that Schwinger was someone very special and should be let alone. Schwinger did not want to be caught in the maelstrom of Oppenheimer's persona. He later remarked, "Oppenheimer was a dominating personality, but I didn't want to be dominated. So to some extent I would back away and say: I have to think about this in my own way." Nonetheless Oppenheimer kept him on a second year as his research associate. He then went to Purdue as an instructor, but when Oppenheimer asked him to come to Los Alamos he declined—again, he did not wish to be "dominated." He made monumental contributions to the development of radar during the war. Schwinger's reaction to Oppenheimer was unusual. Most of his students were devoted to him to the extent of imitating his mannerisms and even his speech patterns.*

Oppenheimer was justly proud of the school he had created. At his hearing, when he presented his "autobiography," he noted that "Starting with a single graduate student in my first year in Berkeley, we gradually began to build up what was to become the largest school in the country of graduate and postdoctoral study in theoretical physics, so that as time went on, we came to have between a dozen and 20 people learning and adding to quantum theory, nuclear physics, relativity and other modern physics. As the number of students increased so in general did their quality; the men [he seems to have forgotten the women] who worked with me during those years hold chairs in many of the great centers of physics in this country; they have made important contributions to science, and in many cases to the atomic-energy project. Many of my students would accompany me to Pasadena in the spring after the Berkeley term was over, so we might continue to work together." This conveys the facts but does not account for them—Oppenheimer's personality. Part of that personality was a taste for risk.

---

*Schwinger's students at Harvard did the same. You could always tell one because they pronounced "nuclear" as "nucular," a la Schwinger.

Perhaps this was a reaction to the overprotectiveness of his parents, but he seemed to relish putting himself in physical danger.

This was true of many of his activities. On his ranch in New Mexico he rode horses. His brother reported that Oppenheimer once decided that, because it often rained during the day, they would ride at night. They rode on narrow mountain trails and, as Frank Oppenheimer noted, it rained anyway. Serber remembered that one night at about midnight they were riding over a high mountain pass in a lightning storm. Coming to a fork in the trail, Oppenheimer said that if they took one branch it would be about seven miles to get home. The other branch, he noted, was a little longer but much more "beautiful."

In California, Oppenheimer was a notoriously dangerous driver. He bought a large Chrysler which he called "Garuda," after the bird deity in Hindu mythology. He was pleased because he could get the car to go ninety-five miles an hour. He enjoyed racing trains. On one occasion he was racing a train and had a crash. The woman passenger who was with him was knocked unconscious but was not seriously injured. Oppenheimer's father gave her a Cézanne drawing and a Vlaminck painting as an apology. It is certainly odd that he felt the need of doing this on behalf of his son. As far as I know, apart from sailing—which he also did dangerously—and riding, Oppenheimer had no interest in sports. On one occasion, when I happened to ride on the commuter train with him from Princeton to New York, he asked me what I was going to do in the city. When I told him I was going to watch a professional tennis match, he looked at me as if I was demented.*

This apparent indifference to danger also manifested itself at Los Alamos. At the Trinity test site there was an arrangement to

---

*Pancho Gonzales was playing.

film the steps that led up to the first nuclear test explosion in July 1945. This was a plutonium bomb. It used high explosives shaped to act like lenses to produce the compression of the plutonium "pit" to the high densities needed to make the nuclear explosion. Placing the "pit" into the explosive assembly was done manually. Working with high explosives must have been extremely dangerous; the film shows several men doing it. Looking over their shoulders is Oppenheimer, who seems quite unconcerned. He is wearing the porkpie hat that had become his trademark, and an open-necked shirt, not his usual suit. It is one of the rare sets of photographs involving him where he is not shown smoking.*

In the decade before the war, Oppenheimer and his various associates did an astounding variety of work. Much of it has by now entered the standard physics textbooks and is no longer attributed unless it is something like the Oppenheimer-Phillips process which bears their name. Let me give a couple of examples. In 1928, Dirac discovered the equation that carries his name. It is an equation that describes an electron—in fact any particle in the same general class as the electron.† Until Dirac's work, the equations that had been proposed for the electron did not include the effects of the theory of relativity. This limited their applicability to electrons that moved at speeds that were not close to the speed of light—a serious defect that Dirac's equation fixed, but at a cost. The cost was that there seemed to be too many solutions. The unwanted solutions had the very unphysical property that they corresponded to electrons with a negative energy. But Dirac noticed

*Roger Meade of Los Alamos pointed out to me that because of the tight space these people were working in, Oppenheimer's presence must have been somewhat of an annoyance.

†Not to be too obscure, these are particles such as the neutron, the proton, and the neutrino, which have an intrinsic angular momentum (spin) in units of Planck's constant of 1/2.

that these unwanted solutions could also be interpreted as particles with positive energies, but with electric charges of the opposite sign. We would now call these "anti-particles." But where were they? The only particle that seemed to fit was the proton. It had a positive charge though a very different mass from the electron. Dirac suggested this interpretation but in 1930, in one of his most important papers, Oppenheimer showed that this was impossible because matter would be unstable. The nuclear protons and the atomic electrons would annihilate each other.* A year later Dirac wrote that, because of Oppenheimer's argument, he was now persuaded that these particles would be positively charged electrons-positrons. He argued that one would not find them in nature because they would annihilate too rapidly with electrons and thus disappear. In this he was too pessimistic. In 1932, C. D. Anderson of the California Institute of Technology found them in cosmic rays. If Anderson had any awareness of the theoretical arguments that preceded his discovery, he did not say. As an afterthought, he suggested looking for a negatively charged proton—the anti-proton. It was found in 1955, using one of the accelerators that had evolved from Lawrence's cyclotron.

To me, the most interesting of these prewar papers were three that Oppenheimer wrote in the late 1930s on what we would now call astrophysics. They were written with different collaborators, but all addressed the general question of the ultimate fate of stars. Although the exact mechanisms that generate stellar energy were not discovered until the end of the 1930s, it was agreed earlier that

---

*A physicist reading this will be puzzled. The process electron plus proton goes into gamma rays is forbidden, at least to a very high order of approximation, by the conservation of baryon number. But Oppenheimer's calculation was done in the context of the "hole" theory, where this process was possible. It seems that, in any event, he overestimated its probability.

it must have to do with nuclear energy. The nuclear processes that produce energy are of two kinds: fission and fusion. Fission, which we will come back to when we discuss atomic bombs, is the splitting of a heavy nucleus, like uranium, into lighter nuclei. The mass difference between the initial and final nuclei is available as energy, in accordance with Einstein's formula $E=mc^2$. Fusion, which we will also come back to when we discuss the hydrogen bomb, is just the opposite. Light nuclei "fuse" into even lighter nuclei, with again an energy release according to Einstein's formula. In the Sun, for example, we start essentially with protons that by a series of fusions and other reactions end up as a helium nucleus. Eventually one runs out of protons, but then the helium nuclei can react to produce carbon and oxygen. The Sun stops at carbon and oxygen. In stars much more massive than the Sun, the reactions continue until iron is produced. (Iron cannot undergo energy-producing nuclear reactions.) This activity goes on in the core of the star. When these reactions stop, the core contracts under the force of gravity. Surrounding it is a gaseous envelope which expands because the nuclear reactions in the core continue to heat it up.

Take a star about as massive as the Sun. With the contraction of the core, the electrons around the atoms are freed and form a kind of gas. But it is a highly peculiar gas. The reason goes back to a theoretical discovery by Pauli. Even before the creation of quantum mechanics in its modern form, Pauli stated what is known as the "exclusion principle." In the case of electrons, it mandates that no two electrons can be in identical states.* This helps us under-

---

*As mentioned in the preceding footnote, electrons have a spin of 1/2, meaning that this spin can point either "up" or "down." These are two different states. We cannot have three electrons with no angular momentum except their spins, because two of them would then be pointing up—or down—whch Pauli forbids.

stand how atoms are built up since it restricts the number of electrons in each shell of electrons that orbit around the nucleus.
These highly compressed stars are known as White Dwarfs—white
because of their low luminosity as compared to normal stars, and
dwarfs because, though they have masses comparable to the Sun,
they have sizes comparable to the Earth! They are so dense that a
teaspoon of a White Dwarf would weigh five tons. The electrons
are packed close together; but Pauli's principle does not allow
them to be in exactly the same state, so they resist the gravitational
squeezing. If these two effects balance, one has a stable stellar
core—that is a White Dwarf. But if the star is too massive, even the
pressure caused by the mutual repulsion of the electrons due to
the Pauli principle is not enough to forestall the gravitational collapse. The question is, how much mass is too much?

This was the problem that was first tackled by the Indian-born
astrophysicist Subramanyan Chandrasekhar. He seems to have
done it in 1930, on board the ship that was bringing him from
India to England, to Cambridge University. We can think about
what Chandrasekhar did in terms of energy. The gravitational attraction produces a negative energy while the Pauli repulsion produces a positive energy. So long as the total energy is positive, a
White Dwarf can be formed. But the limit is when the two energies just balance. If one puts in the mathematics, one gets an expression for the mass at which this happens. Chandrasekhar found
that it was about 1.2 solar masses. Modern calculations give a value
of about 1.4 solar masses. But this still left the question of what happens to stars that are more massive than this.

Since they cannot form White Dwarfs, they continue to collapse under the force of gravity. When the density is high enough,
the electrons combine with the protons in the core to produce
neutrons and neutrinos. The latter are ghostly particles with very

little mass that escape from the core, which is then composed of neutrons—a neutron star. These are some of the strangest objects imaginable. They have masses larger than the Sun but with radii of a few miles! You could fit one in the middle of a large city. The first suggestion that such an object might exist was given by the Russian physicist Lev Landau in 1932. He did the work *before* the neutron was discovered, so he actually considered a collapsed star with the density of an atomic nucleus. In 1934 the astronomers Walter Baade and Fritz Zwicky published one of the most remarkably prescient astronomy papers ever written. They invented the notion of a supernova—the incredibly powerful explosion when certain stars collapse—and suggested that in the detritus of a supernova explosion there might remain what they called a "neutron star," a star with very closely packed neutrons. They were right. In 1967 the English astronomers Jocelyn Bell and Antony Hewish discovered "pulsars"—very rapidly rotating radio sources. They were the neutron stars predicted by Baade and Zwicky. Little of this was understood in detail before Oppenheimer and his students got into the field. Their work resulted in a series of papers that were probably Oppenheimer's most successful scientific work.

The first paper, a brief note on this subject—"The Stability of Stellar Neutron Cores"—was written with Serber. It is basically a critique of Landau's work, especially his 1938 paper which dealt with such neutron cores. Oppenheimer and Serber point out that the physics of these stars is more complex than that of the White Dwarfs because neutrons interact with one another strongly, with a force that is not as well understood as the electromagnetic force with which the electrons in a White Dwarf interact. This set the stage for the next paper, "On Massive Neutron Cores," written with a graduate student, George Volkoff. Volkoff, who died in the year 2000, was an interesting man. He had been born in Russia in

1914 and had emigrated to Canada when he was ten. An adventurer and linguist, he did some remarkable traveling. He helped create the Canadian reactor program. His paper with Oppenheimer was the first to tackle the physics of neutron stars.

In addition to the complexity of the nuclear forces in neutron stars mentioned earlier, there is also a complication with the theory of gravitation. With the White Dwarfs the gravitational fields are sufficiently weak so that, to a good approximation, one can use the gravitational theory of Newton. This you cannot do with neutron stars, as was well understood. Gravitation is too strong, and you must use Einstein's general theory of relativity and gravitation. This immediately complicates matters because the theory is highly complex and the equations notoriously difficult to solve. In fact the relevant equations had never been written down for the neutron star case. Oppenheimer and Volkoff were fortunate to have the assistance of Richard Tolman. Tolman, who was a colleague of Oppenheimer's at Cal Tech—he used to stay with the Tolmans when he taught there—was one of the acknowledged masters of the theory of relativity. Oppenheimer and Volkoff were able to lean heavily on Tolman's work and his advice.

Of course, once you write down the equations you have to try to solve them. Oppenheimer and Volkoff had to do this numerically, one supposes with slide rules. This enabled them to derive the conditions under which these stars would be stable. They found an upper limit to the mass—the Oppenheimer-Volkoff limit. Above this mass a neutron star would continue to contract under gravitation. The limit they proposed was about 0.7 solar masses. Present estimates are between 3 and 5 solar masses. It is not a very certain number because of all the difficulties in the calculation. Oppenheimer and Volkoff concluded pessimistically that "it seems unlikely that static neutron cores can play any great part in stellar evolution; and the question of what happens after energy sources are exhausted, to stars greater than 1.5 [solar masses] still remains unanswered."

In the next paper in the series—one of the great papers in twentieth-century physics—a possible answer is given. Entitled "On Continued Gravitational Attraction," this paper was written with another brilliant graduate student, Hartland Snyder. Snyder later went on to become one of the inventors of the technique by which beams of particles are kept together in all the modern particle accelerators, machines that have replaced the old cyclotrons. It is an essential ingredient of their design. In his book, Serber remarks that Snyder, who he says was from Utah, and had apparently driven a truck, was the best mathematician in their group. In the 1939 paper, Oppenheimer and Snyder were able to make use of a development in Einstein's theory that took place in 1916, just after Einstein published his basic paper. It was done by the German astronomer Karl Schwarzschild who, the same year, died from a skin disease that he may have contracted while serving on the Russian front. He was forty-two. Schwarzschild was interested in how gravity would distort space and time outside of a spherical distribution of mass. This would determine, say, planetary orbits around such a mass. He did not consider what would happen if this mass began to contract. In fact, no one had considered that. One of the reasons may have been—in Einstein's case it was—the fact that at a certain distance from the center of the mass (a distance known as the Schwarzschild radius) the equations appear to go crazy. Terms become infinite. Since a collapsing mass will shrink below this radius eventually, it looks as if one has reached some sort of stalemate. Ironically, at about the same time Oppenheimer and Snyder were doing their work in California, Einstein in Princeton was attempting to show that the theory had an incurable disease: the singularity at the Schwarzschild radius. He showed that no stable star could exist with this radius. His reasoning was correct but irrelevant. Oppenheimer and Snyder, who were unaware of it, were studying stars that were not stable but were collapsing.

To treat this new case in which stars were collapsing, they had to derive the equations—which they were able to do—and then solve them, which they could not. Nonetheless by using the features of the equations they were able to tell how the solutions would behave. This involved some ingenious mathematics. They studied two observers with different viewpoints. First they asked how the collapse would appear to an observer far away from the collapsing mass. They imagined that a clock could be attached to the surface of this mass. This outside observer could observe how the clock behaved when compared to the clock that he had at hand. It is well known that a clock in a gravitational field runs more slowly. In this case, as the collapsing mass approaches the Schwarzschild radius, it slows down to a dead stop as observed from the outside. It takes, according to the outside observer, an infinite amount of time to collapse to this radius, so what happens when it gets there is irrelevant to him, since it never does.

To an inside observer moving with the mass, the situation appears very different. This observer sees a rapid collapse to the Schwarzschild radius and beyond. Nothing strange happens at this radius except that the gravitational forces crush him.

Now for the most interesting thing that Oppenheimer and Snyder found. Suppose that this inside observer tries to explain to the outside observer what is going on by means of, say, light signals. He faces an ever-diminishing part of the surface from which the light can escape to reach the outside observer. At the Schwarzschild radius there is no part of the surface where this can happen, and the inside observer is cut off from the rest of the universe.* All of this is spelled out clearly in the Oppenheimer-Snyder paper.

In 1952 the Princeton physicist John Wheeler agreed to give a course on relativity and gravitation, a subject he had never studied. He came upon the Oppenheimer-Snyder paper and was so transfixed that he made it his field of research. Sometime later, during a lecture, someone suggested to Wheeler the name "black hole" for this collapsing star, and it has been with us ever since. Wheeler told me that in the 1960s, not long before Oppenheimer's death, he tried to talk with him about this work. Oppenheimer had absolutely no interest. The subject had gone out of fashion. If he had lived a few more years, he would have seen it blossom into one of the centerpieces of contemporary physics.

DURING THESE California years Oppenheimer had a full social life with many friends of every kind. But three of them are of particular importance. The relationships he had with these people all ended badly and were highly destructive. They were Ernest Lawrence, Jean Tatlock, and Haakon Chevalier. Here are some preliminaries about them, beginning with Lawrence whom we have already met. They will reenter our story in the chapters that follow.

Lawrence and Oppenheimer were antipodes. Lawrence's grandparents had come from Norway. He was born in 1901, in Canton, South Dakota. His father was a not especially well-paid

---

*This is not really true. Stephen Hawking showed that some radiation is emitted—"Hawking radiation"—but this is an effect of quantum mechanics which Oppenheimer and Snyder were not considering.

superintendent of the Northern Normal and Industrial School in Aberdeen, South Dakota. The family was Lutheran. Lawrence earned his way through St. Olaf College by selling pots and pans door to door. He earned his Ph.D. from Yale in 1925 and stayed there as an assistant professor for three more years, during which time he was not very happy. Berkeley was in the process of trying to build its physics department, and Lawrence's name surfaced as an up-and-coming physicist who might help. They offered him an associate professorship, which was a promotion. He went to Berkeley in 1928, a year before Oppenheimer, and stayed there the rest of his life.

Being a bachelor, Lawrence moved into the Faculty Club where he was joined a year later by Oppenheimer. The two men became very close friends. It seems as if they even double-dated. Lawrence made visits to Perro Caliente where horseback riding was obligatory. One of my favorite photographs of Oppenheimer was taken at the ranch. It shows Oppenheimer leaning against a large black roadster. (This must have been "Garuda.") His hair is very long. He is wearing an elegant-looking shirt, open at the neck, with a cardigan over it. His pants, with the bottoms inside a pair of cowboy boots, have a fashionable cut. They are held up by a belt with a large silver buckle. He looks like someone who would fit in nicely at a fancy country club. Lawrence, on the other hand, looks a bit like a country bumpkin. He is wearing a shirt and tie with a sweater over it and a somewhat ill-fitting jacket—decorated in squares—over that. He has on a pair of jodhpurs with formal riding boots. It looks as if his stomach is sticking out. But he has a broad, self-confident smile. My guess is that Lawrence took Oppenheimer's antics with a large dose of salt. Nonetheless the two of them had a great affection and respect for each other.

When, in October 1931, Oppenheimer's mother was dying of

leukemia, and he was in New York to be with her, Lawrence sent roses from California. Oppenheimer responded,

> Dear Ernest,
>
> Your sweet message, and the lovely roses, came last night, as my father and I were sitting together after dinner. Both of us were very much touched, Ernest, and grateful to you. My father asked me to tell you from him his appreciation and his thankfulness; several times during the evening he said, "That was such a fine thing of Lawrence to do."
>
> Things are pretty bad here, now. Mother, after a short respite, has been growing rapidly very much worse; she is comatose, now; and death is very near. We cannot help feeling now a little grateful that she should not have to suffer more, that she should not know the despair and misery of a long hopeless illness. She has always been hopeful and serene; and the last thing she said to me was "Yes—California." But even at that it is not very easy. . . .

This letter was written on October 16, 1931. The following day Oppenheimer's mother died. As she was dying, Oppenheimer said to a friend, "I am the loneliest man in the world." Oppenheimer always had a tendency to dramatize things. Nonetheless, for a man approaching thirty to say something like this at the death of his mother is certainly revealing. In 1932, Lawrence married Mary "Molly" Blumer, the daughter of a Yale professor. At first this did not change Oppenheimer's relationship with Lawrence; he simply fit in as one of the family. But in the next ten years Lawrence and his wife came to loathe Oppenheimer, both for what they presumed to be his morals and what they knew were his politics.

As I have mentioned, until the mid-1930s Oppenheimer was totally apolitical. This is one of the things that most struck Rabi when he met him. To Rabi he was an unworldly aesthete. This is how he appeared to everyone. But all of this changed dramatically in 1936. In his "autobiography" prepared for his 1954 security hearing, Oppenheimer wrote,

"Beginning in late 1936, my interests began to change. These changes did not alter my earlier friendships [they certainly altered his friendship with Lawrence], my relations to my students or my devotion to physics; but they added something new. I can discern in retrospect more than one reason for these changes. I had a continuing, smoldering fury about the treatment of Jews in Germany. I had relatives there, and was later to help in extracting them and bringing them to this country. I saw what the depression was doing to my students. Often they could get no jobs, or jobs which were wholly inadequate. And through them I began to understand how deeply political and economic events could affect men's lives. I began to feel the need to participate more fully in the life of the community. But I had no framework of political conviction or experience to give me perspective in these matters."

He goes on,

"In the spring of 1936, I had been introduced by friends to Jean Tatlock. Her father was a noted professor of English at the university [John Tatlock was a specialist in Chaucer and medieval literature in general]; and in the autumn I began to court her, and we grew close to each other. We were at least twice close enough to marriage to think of ourselves as engaged. Between 1939 and her death in 1944 I saw her very rarely. She told me about her Communist Party memberships; they were on again, off again affairs, and never seemed to provide for her what she was seeking. I do not believe that her interests were really political. She loved this country and its people and its life. She was, as it turned out, a friend of many fellow travelers and Communists, with a number of whom I was later to become acquainted."

When Oppenheimer met her, Jean Tatlock was in her mid-twenties and working on a graduate degree in medicine at Stanford. She was, as Oppenheimer says, deeply involved with left-wing activities such as Spanish Civil War benefits. Indeed, Oppenheimer seems to have met her at such a benefit. One of the Communist friends Tatlock introduced him to was the poet Edith Jenkins. In her book *Against a Field Sinister*, she describes Tatlock and Oppenheimer. She writes,

"She was private, too, about her relations with Oppie. All of us were a bit envious. I for one had admired him from a distance. His precocity and brilliance already legend, he walked his jerky walk, feet turned out, a Jewish Pan with his blue eyes and his wild Einstein hair. And when we came to know him at the parties for Loyalist Spain, we knew how those eyes would hold one's own, how he would listen as few others listen and punctuate his attentiveness with 'Yes! Yes! Yes!' and how when he was deep in thought he would pace so that all the young physicists who surrounded him walked the same jerky, pronated walk and punctuated their listening with 'Yes! Yes! Yes!'"

Not all their mutual acquaintances were equally impressed. Jenkins tells of a party where Oppenheimer had delivered an epigram. An elderly listener, "deep in his cups and in the solitary wisdom they allowed him, mumbled, 'Never since the Greek tragedies has there been heard the unrelieved pomposity of a Robert Oppenheimer.' With that, he laid his head upon the table, a Tenniel dormouse, fast asleep."

What exactly Oppenheimer's relationship to Tatlock was, we

cannot be sure. From what we do know, it seems to have been somewhat one-sided. In an interview, Serber recalled that "she disappeared for weeks, months sometimes, and then would taunt Robert mercilessly. She would taunt him about who she had been with and what they had been doing. She seemed determined to hurt him, perhaps because she knew Robert loved her so much." Their affair ended in 1939. That August, Oppenheimer met Kathryn "Kitty" Puening at a garden party at the Tolmans' in Pasadena. Kitty had been born in Germany in 1910. Her mother claimed to be a distant relation of General Wilhelm Keitel, the Nazi general to whom Hitler gave the responsibility of hunting down the participants in the 1944 assassination plot. He was tried at Nuremberg and hanged. Her mother had been briefly engaged to him but married someone else. Her parents—her father was a mining engineer—moved to Pittsburgh when she was two. Serber had met her in Philadelphia in 1938, where she was studying biology at the University of Pennsylvania. She was then engaged to be married to a British doctor, Stewart Harrison, whom she had met in England. She had earlier been married to Joe Dallet, a Communist organizer from Ohio. He had joined the Abraham Lincoln Brigade which fought in Spain, and had been killed in the Battle at Saragossa in 1937. She herself had joined the Communist party in 1934, after she met Dallet, but had left it. When Oppenheimer met her in 1940 she was still married to Harrison. Yet the two of them began an affair that scandalized some of the academic community. A divorce followed quickly and when she married Oppenheimer in Las Vegas in November 1940, she was already pregnant.

After the marriage, Jean Tatlock visited Edith Jenkins, who wrote: "Jean stayed with us for about a week in our San Francisco apartment on Buena Vista East. She enjoyed holding Margy [Jenkins's baby daughter] in her arms, looking out the window at the green park across the way. Dave [Jenkins's husband] had not yet come home, and we were having the kind of intimate discussion we had not often had in the days when we were really more intimate. Oppie was already married to Kitty. I asked Jean if she regretted refusing to marry him, and she said, yes, she did not think she would have done so had she not been so mixed up. I recall responding perhaps it was that she perceived him as essentially nonsexual. Jean put her cheek against Margy's and held gently the baby hand that was pulling her hair, and said 'Maybe you're right. I wish I could meet a man like Dave.'"

My guess is that until Oppenheimer met Kitty he had never had a fully satisfactory sexual relationship with a woman.

At his hearing, Oppenheimer was extensively questioned about his relationship with Tatlock. At one point there was a colloquy which is poignant both for what it says and what it doesn't say. Oppenheimer is designated as "A" and Roger Robb, the attorney questioning him, as "Q." The dialogue is long, but one cannot get a flavor of what the proceeding was like without quoting it. It begins with Robb.

Q. . . . Between 1939 and 1944, as I understand it, your acquaintance with Miss Tatlock was fairly casual; is that right?

A. Our meetings were rare. I do not think it would be right to say that our acquaintance was casual. We had been very much involved with one another and there was still very deep feeling when we saw each other.

Q. How many times would you say you saw her between 1939 and 1944?

A. That is 5 years. Would 10 times be a good guess?

It may seem puzzling that Oppenheimer would put this as a question. One must keep in mind that once he began working on nuclear weapons, he was shadowed by the FBI night and day. During the hearing he was constantly confronted with information about himself that he had forgotten in detail. The dialogue goes on:

Q. What were the occasions for your seeing her?
A. Of course, sometimes we saw each other socially with other people. I remember visiting her around New Year's of 1941.
Q. Where?
A. I went to her home or to the hospital. I don't know which, and we went out for a drink at the Top of the Mark. I remember that she came more than once to visit our home in Berkeley.
Q. You and Mrs. Oppenheimer.
A. Right. Her father lived around the corner not far from us in Berkeley. I visited her there once. I visited her, as I think I said earlier. In June or July of 1943.
Q. I believe you said in connection with that you had to see her.
A. Yes.
Q. Why did you have to see her?
A. She had indicated a great desire to see me before we left [for Los Alamos]. At that time I couldn't go. For one thing, I wasn't supposed to say where we were going or anything. I felt she had to see me. She was undergoing psychiatric treatment. She was extremely unhappy.

The chronology here is a bit confusing. Los Alamos got started in earnest in April 1943. The reference to "at that time" must refer to an earlier date. The "June or July" dates refer to a trip to Berkeley that Oppenheimer made after he was already at Los Alamos. It was June 13 when he met Tatlock. "I felt she had to see me" evidently refers to the latter date. That he contacted Tatlock then, even in response to her request, was certainly indiscreet. Robb now picks up this thread.

Q. Did you find out why she had to see you?

A. Because she was still in love with me.

Q. Where did you see her?

A. At her home.

Q. Where was that?

A. On Telegraph Hill.

Q. When did you see her after that?

A. She took me to the airport, and I never saw her again.

Q. That was in 1943?

A. Yes.

Q. Was she a Communist at that time?

A. We didn't even talk about it. I doubt it.

Q. You have said in your answer that you knew she had been a Communist?

A. Yes. I knew that in the fall of 1937.

Q. Was there any reason for you to believe that she wasn't still a Communist in 1943?

A. No.

Q. Pardon?

A. There wasn't, except that I have stated in general terms what I thought and think of her relation with the Communist Party. I do not know what she was doing in 1943.

Q. You have no reason to believe she wasn't a Communist, do you?

A. No.

Q. You spent the night with her, didn't you?

A. Yes.

Q. Did you think that consistent with good security?

A. It was, as a matter of fact. Not a word—it was not good practice.

Q. Didn't you think that put you in a rather difficult position had she been the kind of Communist that you have described her or talk about this morning?

A. Oh, but she wasn't.

Q. How did you know?

A. I knew her.

One can feel the agony of this exchange. What the transcript does not tell us is that some six months later, January 5, 1944, Jean Tatlock committed suicide. She filled the bath in her apartment, took a number of sleeping pills, and then drowned herself. Some days later her father discovered the body. When Oppenheimer heard, he was devastated. Her friend Edith Jenkins had a somewhat different view. She recalled the poetry that she and Tatlock had shared. They both had discovered Donne at about the same time, and Tatlock introduced his poetry to Oppenheimer. Donne became one of his favorite poets, someone to read in dark times. In Jenkins's memoir, she writes,

"For all the confidential exchange, I felt somehow remote from her, however. I thought it was the fact that she was now a Freudian analyst and we considered Freud and Marx unreconcilable, though she claimed she was still a Marxist. But I did not feel close, and when her suicide note appeared in the paper, I could not grieve. I thought perhaps it was that I knew she wanted to die. And besides, her note did not let us mourn, because it said, 'To those who loved and helped me' (not 'who tried to help me'—how considerately she chose her words even then)—'all love and courage.'"

Then, later in her memoir she writes,

"Jean was discovered by her father. The door of the Telegraph Hill apartment was bolted so that J. S. P. Tatlock, who was old then and Emeritus Professor of Chaucer at Berkeley, climbed through a window and found her. She had taken a lot of sleeping pills and was lying in the bath tub. I cannot get the picture out of my mind how her large breasts must have floated in the water. He had no right to see her that way. Her note had said: 'all love and courage.' She had not in any way abandoned hope except, that is, for herself."

The third member of this trio of Oppenheimer friends is Haakon Chevalier. He will reenter our story when we discuss Oppenheimer's hearing, where he plays a central role. Here I want to give the background. Chevalier had been born in Lakewood, New Jersey, of French and Norwegian parents. He had spent some time at sea going around the world on a schooner before beginning an academic career. When Oppenheimer met him in 1937 he was thirty-five, two years older than Oppenheimer. He had become a professor of French literature at Berkeley and had translated André Malraux and written an important book on Anatole France. He later wrote a book about his friendship with Oppenheimer—*Oppenheimer: The Story of a Friendship*. By the time he wrote it in 1965, it was a friendship that had gone terribly wrong, but it nonetheless retains a sense of the hero worship he felt. He recalls the Oppenheimer he first knew:

"I can remember my earliest impressions of him at this stage. He was tall, nervous and intent, and moved with an odd gait, a kind of jog, with a great deal of swinging with his limbs, his head always a little to one side, one shoulder higher than the other. But it was the head that was most striking: the halo of wispy black curly hair, the fine sharp nose, and especially the eyes, surprisingly blue, having a strange depth and intensity, and yet expressive of candor, that was altogether disarming. He looked like a young Einstein, and at the same time like an overgrown choir-boy. There was something both subtly wise and terribly innocent about his face. It was an extraordinarily sensitive face, which seemed capable of registering and conveying every shade of emotion. I associated it with the faces of apostles, either imagined or remembered from Renaissance paintings. A kind of light shone from it, which illuminated the scene around him."

Chevalier was a member of the Communist party, and one of the things that cemented his friendship with Oppenheimer was the left-wing causes that interested both of them. Chevalier and Oppenheimer founded the Berkeley chapter of the American Federation of Teachers, which was largely devoted to politics of a nonacademic sort. Oppenheimer began giving relatively large sums to a wide variety of organizations, many of which were certainly Communist fronts. When asked about this in a 1943 security interview, he said to the security officer who was questioning him that he had probably belonged to every Communist front organization on the West Coast—a surprisingly careless statement under the circumstances. He was quizzed on this at his hearing and said that he had meant it as "half jocular overstatement." In this instance his thoughtlessness had only hurt himself, but in some of his other statements he hurt other people—among them Chevalier—and very badly. Given Oppenheimer's genius and general sensitivity, I wondered why, so I asked Rabi, who knew him as well as anyone. This is what he said:

"You know Oppenheimer. Once he gets into something he gets into it with both feet. He becomes a leader. He was like a spider with this communication web all around him. I was once in Berkeley and said to a couple of his students, 'I see you have your genius costumes on.' By the next day Oppenheimer knew that I had said that. He was practically running the local teachers' union. Pauli once said to me that Oppenheimer was a psychiatrist by vocation and a physicist by avocation. He had this mystic streak that could sometimes be very foolish. Sometimes he made foolish judgments and sometimes he just liked to tell tall stories. He was a very adaptable fellow. When he was riding high he could be very arrogant. When things went against him he could play the victim. He was a most remarkable fellow."

We shall see later how this affected Chevalier.

In 1942, Oppenheimer published his last physics paper before the war—"Pair Theory of Meson Scattering." He had already begun thinking about nuclear weapons and was about to make the most fundamental adaptation of his life.

*Oppenheimer with General Groves at the base of the steel tower
on which the first atomic bomb hung when tested near
Alamogordo, New Mexico, 1945*

3

LOS ALAMOS

We were lying next to each other on the desert near the control hut. . . . There was a sense of this ominous cloud. It was a very scary time. I think we just said, "It worked." No one knew if it would work.

—Frank Oppenheimer, describing how he and his brother watched the first test explosion at Alamogordo on July 16, 1945

IN NOVEMBER 1942, Stirling Colgate was a pupil in the Los Alamos Ranch School. One of his schoolmates was Gore Vidal. The school was designed for upper-middle-class boys. They lived fairly rough, but each of them had a horse which they were responsible for. Colgate was interested in physics even then. He went on to become a leading nuclear weapons physicist who spent the latter part of his career at Los Alamos. On November 16 four men appeared at the school in civilian clothes. Colgate thought he recognized one of them from a picture in his physics text. He thought it was Ernest Lawrence; it was actually Lawrence's associate, Edwin McMillan. Lawrence and McMillan had often been photographed together at one of the cyclotrons in Berkeley, so the confusion was plausible. Colgate did not recognize Oppenheimer, nor General Leslie Groves and Major John Dudley who had been given the job of trying to locate a site for a new nuclear weapons laboratory. Groves had rejected Dudley's first choice, a canyon forty miles from Santa Fe. Oppenheimer knew this country very well since it was not far from his ranch, and he suggested Los Alamos, a site that Dudley had already visited but did not like. When Groves saw the mesa at Los Alamos he decided, at once, that this would be the location of the laboratory. This was three years after the Hungarian-born physicist Leo Szilard was able to deliver Einstein's letter urging a nuclear research program (which Szilard had drafted) to Alexander Sachs, an economist who had ties with President Roosevelt, to whom the letter was addressed.

Many people seem to think that this 1939 letter, which warned of a German nuclear program, is what started the atomic bomb project that finally led to Los Alamos in the spring of 1943. But the U.S. program was in fact stalled for several years, so that if one needs a starting date after which the bomb project was taken seriously, my own choice would be May 19, 1940. It was on this date that two brief reports written by the physicists Otto Robert Frisch and Rudolf Peierls landed on the desk of Henry Tizard, who was then an important figure in British wartime science. Frisch was a Viennese who had taken refuge in Bohr's institute in Copenhagen. He and his aunt, Lise Meitner, who had emigrated from Germany to Sweden, were the first people to understand that when, in late 1938, the German physical chemists Otto Hahn and Fritz Strassmann bombarded uranium with slow neutrons, they had actually fissioned—split—the nucleus. Peierls was a German-born theorist who, like Frisch, was Jewish and had also been forced to emigrate. The two of them found themselves at Birmingham University in England, unable to participate directly in the war effort because they were technically enemy aliens. At the university was a radar project run by a physicist named Mark Oliphant. Peierls told me that, officially, he was not allowed to know anything about it, but periodically Oliphant would come to him and say, "By the way, I just ran into this interesting electromagnetic problem. I wonder if you have any ideas."

Frisch and Peierls began investigating the prospects for nuclear weapons on their own. Perhaps they were worried that the Germans might be doing the same, though Peierls told me that if the Germans were, he was sure it was not on a large scale. He had somehow managed to get hold of catalogues from the universities where he knew his former colleagues were teaching. He noticed that they were still teaching the same courses at the same places and reasoned that therefore they could not be involved in a major crash project. They both knew that, in 1939, Bohr had shown that only the rather rare* isotope of uranium—U(235)—is fissionable by neutrons of all energies. The common isotope—U(238)—is fissionable only by energetic neutrons. This was crucial for bomb manufacture since one wanted a chain reaction with the neutrons emitted in fission in turn fissioning other uranium nuclei as rapidly as possible. To bring this about one needed to separate the isotopes to get rid of the abundant U(238). In fact, before Peierls arrived in Birmingham, Frisch had already begun investigating how this separation might be carried out. Bohr himself had decided that it was so difficult as to rule out any realistic program to make nuclear weapons. This was a position he held until the summer of

---

*In natural uranium, U(235) occurs in only about .7 percent of any sample.

1943, when he became convinced, erroneously, that the Germans were making serious progress in creating a nuclear weapon.

Frisch made the first steps toward the work that led to the papers he and Peierls would eventually write. He actually tried to separate the uranium isotopes. Since, as a bachelor, he had been given a room in the Peierlses' house—the first of many well-known physicists—it was natural to discuss the matter with Peierls. They proceeded on a "what if" basis. What if you could actually separate U(235)? How much would you need to make an explosive device? And what would be the consequences? They wrote up the results of this investigation in two short reports: "Memorandum on the Properties of a Radioactive 'Super-bomb'" and "On the Construction of a 'Super-bomb'; Based on a Nuclear Chain Reaction in Uranium.*

Two of the most important sentences in their first paper are: "It is a property of these super-bombs that there exists a 'critical size' of about one pound. A quantity of separated uranium isotope that exceeds the critical amount is explosive; yet a quantity less than the critical amount is absolutely safe." In computing this "critical mass," Peierls made some optimistic assumptions. Depending on the configuration, the actual mass required is at least an order of magnitude larger than a pound—but we are not talking about tons! It was this optimistic estimate that landed on Tizard's desk.

In the second report Frisch and Peierls go into details. They even propose a design consisting of two parts that, put together, form a sphere. Each part is subcritical, but when brought together explosively the chain reaction begins. They noted that the assembled sphere would have a radius of less than about three centime-

---

*The word "super" here should not be confused with the term that was used to designate the early hydrogen bomb designs. We come to this later.

ters—smaller than a tennis ball. They estimated that a bomb of this size would have the explosive power of several thousand tons of TNT.

What is impressive about these papers is their absolute clarity. I would claim that there is more clear understanding of the physics of nuclear weapons in these two short papers than the Germans exhibited in the five years their program lasted. It has made me wonder that, if we had had a real understanding of the German program and its deficiencies at the time, would there ever have been a Los Alamos—and a Hiroshima?

Frisch and Peierls presented their papers to Oliphant, who in turn gave them to Tizard. It is very unlikely that Tizard had ever heard of the two physicists. Tizard proposed that a small committee be formed to study the matter. This eventually became the MAUD committee. The name arose when the British received a message from Bohr in which he asked to convey his greetings to a Maud Ray of Kent. The assumption was that this was an anagram for something connected with nuclear bombs. It turned out there *was* a Maud Ray of Kent who had taught the Bohr children English. Eventually there were contacts between the British and American scientists. But when Tizard came to the United States in 1940 with the British secrets on radar, he found the bomb project to be a very low priority in the States. In fact, when James Bryant Conant—the president of Harvard who would play a major role in the nuclear weapons program—came to England in the late winter of 1941, he had never heard of it until the British told him about it. Conant later credited Oliphant with the spark that set the American program going, a spark that had been ignited by Frisch and Peierls.

By the spring of 1941, Lawrence and his colleagues at Berkeley had begun to make progress in separating minute amounts of

U(235) and in manufacturing microscopic amounts of plutonium, which they had discovered fissioned at least as well as uranium. By that fall, Oppenheimer, who was at Berkeley and in contact with Lawrence, had begun his work on nuclear weapons. Ever since the defeat of France in June 1940, Oppenheimer had been determined to get into the war. His political activities—what Lawrence called his "leftwandering"—had come to a rapid end. This had been a source of major friction with Lawrence, especially when Oppenheimer's political activities were taken into Lawrence's laboratory, which he put a stop to. This had affected their friendship, and the tensions were palpable. Nonetheless it was Lawrence who got Oppenheimer involved in the nuclear program. He arranged for Oppenheimer to be invited to a conference in Schenectady, New York, in October 1941, where fast neutron reactions—the kind that can lead to nuclear explosions—were being discussed. At this conference Oppenheimer presented his own estimate of the critical mass of uranium. I have never been able to find the details of this calculation, so I do not know how it compares with the methods Frisch and Peierls used. Oppenheimer certainly had not heard of their work. He came up with an answer of about five kilograms. During the war, doing and redoing these calculations was a major project. Because of the odd, nonspherical shapes involved, the critical masses had eventually to be determined experimentally. This was done by adding pieces to the assembly until it finally went critical—something that the physicist Richard Feynman called "tickling the dragon's tail." It was so delicate a procedure that the people who added the last masses had to jump off the top of the assembled masses since their bodies reflected neutrons back into it which would have started an explosive chain reaction.

Oppenheimer became associated with a theoretical group led by a University of Wisconsin physicist named Gregory Breit. Breit was an excellent physicist but, to put it mildly, a very difficult personality. He decided that security was being breached somehow in the project and, in May 1942, he quit. Oppenheimer took over as leader. He called himself the "Co-ordinator of Rapid Rupture." One of the first things he did was to call for his old Berkeley postdoctoral Robert Serber to come and help him. In his book Serber recalls,

"The day after arriving in Berkeley, I went down to Oppie's office in Le Conte Hall where he had accumulated a number of British documents concerning bomb design. I remember there was a paper on critical mass and something on assembling the pieces of a bomb. Perhaps there was something on efficiency. I don't remember. [There was. Frisch and Peierls had attempted to estimate it. In an actual bomb only about 2 percent of, say, the uranium is fissioned before the chain reaction is shut off by the expansion of the material. The uranium nuclei get too far apart for the neutrons to reach them before they escape the material.] The papers were rudimentary but were really quite helpful in getting us started."

By July 1942, after Oppenheimer and Serber had been working for a couple of months, Oppenheimer decided to call together a small group of theorists—nine in all—to discuss bomb physics in Berkeley. Among them were Edward Teller and Hans Bethe. Teller was at the University of Chicago, working on fission-related problems. Bethe, who was a professor at Cornell, had been working on radar but was regarded as one of the world's leading nuclear physicists. Oppenheimer called him and managed over an open phone to convey enough of what was going on so that, out of curiosity, Bethe decided to accept his invitation. On the way to California, Bethe stopped off in Chicago to pick up Teller and to have a look at the progress Enrico Fermi was having building a nuclear reactor. The reactor, which was being constructed on a squash court under Stagg Field at the university, did not go critical until December 2 that year, but it was far enough along that Bethe was persuaded it would work and that a nuclear weapon was a real possibility. In an interview, Bethe told me, speaking of himself and Teller,

"We had a compartment on the train to California, so we could talk freely. Teller told me about the idea of making plutonium in the reactor and using the plutonium in a nuclear weapon." In reactors, most of the uranium is in the form of $U(238)$. One of the things a neutron can do when it is absorbed by a $U(238)$ nucleus is to transform that nucleus into one of neptunium, which in turn decays into plutonium.

Bethe went on,

"Teller told me that the fission bomb was all well and good and, essentially, was now a sure thing. In reality, the work had hardly begun. Teller liked to jump to conclusions. He said that what we really should think about was the possibility of igniting deuterium by a fission weapon—the hydrogen bomb. Well, the whole thing was far more difficult than we thought then. About three-quarters of our time that summer was occupied with thinking about the possibility of a hydrogen super-weapon. We encountered one difficulty after another, and came up with one solution after another—but the difficulties were clearly in the majority. My wife knew vaguely what we were talking about, and on a walk in the mountains in Yosemite National Park she asked me carefully whether I really wanted to continue to work on this. Finally I decided to do it. It was clear that the super-bomb, especially, was a terrible thing. But the fission bomb had to be done, because the Germans were presumably doing it."

One of the charges leveled against Oppenheimer at his 1954

hearing was that he was not sufficiently "enthusiastic" about the hydrogen bomb. I will return to this when I discuss the hearing, but it is important to understand which hydrogen bomb Oppenheimer had misgivings about, and why and when. What Oppenheimer and his group were considering in 1942 differed from the invention of Teller and the Polish-born mathematician Stanislaw Ulam in February 1951, which made it clear that a hydrogen bomb could be built. This explanation belongs in the context of the hearings. Here I will describe what preceded the 1951 invention — at least schematically. The reader can be assured that I am not giving away classified information, since I don't have any.

In fission what happens is that a heavy nucleus is split into two medium-weight nuclei and some neutrons — somewhere between two and three. In the experiment of Hahn and Strassmann, for example, a U(235) nucleus was split into barium and krypton along with neutrons. If one compares the masses of the U(235) nucleus and the neutron that initiates the process to the masses produced, one finds that the final masses are smaller. This mass difference is related to an energy by Einstein's formula $E=mc^2$, where m is that mass difference. This energy goes mainly into the kinetic energy of the fission fragments.

Fusion, as we mentioned earlier, works at the opposite end of the nuclear mass scale. Two light nuclei are fused together to produce nuclei that have less total mass. The difference is once again released as kinetic energy. A typical fusion reaction might involve two deuterium nuclei — deuterons — fusing to make a light isotope of helium with a free neutron also produced. The inhibiting factors to fusion are the positive electric charges of the two deuterons. They repel each other electrically and prevent the fusion, which can occur only when the deuterons are close together. But quantum mechanics provides a mechanism for overcoming this—barrier penetration, or tunneling. For this mechanism to be effective, the fusing particles should be as energetic as possible—that is, they should be part of a high-temperature system. Such systems are provided by stars like our Sun. In these stars the material is compressed and heated by the force of gravitation to temperatures where fusion can take place on a large scale.

What the people at Oppenheimer's 1942 meeting were trying to do was to find a way of heating the light elements to a point where fusion would take over. Here is where the fission bomb came in. It seemed to generate enough energy to do the trick—thus the so-called classical super. The canonical classical super consisted—schematically—of a tank filled with liquid deuterium with a bomb inside it. The exploding bomb was meant to "ignite" the liquid deuterium. Apart from the practical problems posed by liquefying deuterium, there was a matter of principle—the laws of physics. No matter what configuration was tried, the heat from the bomb dissipated too rapidly to ignite the deuterium or any of the other light elements that were tried. It was like dropping a match into a gas tank and watching the match go out before it causes the gasoline to explode. From 1942 to 1951, Teller—and anyone he could persuade to work with him—tried to make the thing work, without success. Here, as I have said, is where things stood until the critical breakthrough known as the Ulam-Teller idea.

The person who had the closest contact with President Roosevelt on nuclear matters was an engineer named Vannevar Bush. He had persuaded the president to set up a National Research Defense Committee, which he then headed. In October 1941 he met with the president to brief him about the British report of the MAUD committee. He then won approval to launch a full-scale bomb program in the United States. After some months it became clear that it had grown beyond the scope of various projects scattered around the country, such as those in Berkeley, Chicago, and elsewhere, and that what was needed was a central organization run by the War Department—the army—which would have the authority to set priorities and make all the decisions. One of the weaknesses in the German program was that it lacked this structure. Research was being carried out in twenty-odd institutions located all over Germany and Austria, working often without adequate communication and sometimes at cross-purposes. It devolved onto Secretary of War Henry Stimson to name someone to preside over the whole enterprise, and in September 1942 he selected then Colonel Leslie Richard Groves. Groves, who had been born in 1896 and was a West Point graduate and a professional soldier in the Corps of Engineers, was not pleased. He wanted to be a fighting soldier. Indeed, on hearing the news he said in reference to the bomb, "Oh, that thing. . . ." He managed to get himself promoted to brigadier general before he took the job.

To the best of my knowledge, at the time of his appointment Groves knew nothing about nuclear physics. What he did know was how to manage large-scale construction projects such as the building of the Pentagon, which he had been responsible for. He was also a very quick study. Although he had known about the bomb project only since June, two days after he had been put in charge Groves was in the process of acquiring a site for the separation of uranium at Oak Ridge, Tennessee. What I find remarkable about this is that, at the time, the critical masses were not very well known, so it was not clear how much U(235) would be needed to make the bomb. It was also not clear what the best method for separating the isotopes was, so Groves simply went ahead with all of them. He had a mandate to produce a nuclear weapon as soon as possible, and that is what he was determined to do.

Groves also decided that a separate laboratory was needed

where all the ingredients could be put together. For that he needed a director. Lawrence was an obvious choice. He had been involved in large scientific and engineering projects—his cyclotrons—and he had been working on isotope separation at Berkeley. Furthermore he had a Nobel Prize. One often reads that Groves did not select him because he could not be spared from the work he was doing at Berkeley. My guess is that this was not the real reason. Groves had an uncanny ability to judge people in terms of what role they might play in any project he wanted to get accomplished. He had a gigantic ego and insisted that things be done his way. My hunch is that he saw in Lawrence someone who had already developed a management style, who also had a very large ego, and who might have his own ideas of how to go about things. Lawrence pushed very hard for his associate McMillan, but by that time Groves had decided on Oppenheimer, whom he had met in Berkeley.

Viewed from the outside, Oppenheimer was the most unlikely choice imaginable. He had never managed anything. He was a theorist whose attempts at experimental physics had been disastrous. He was an aesthete who read poetry in several languages, and he had a ton of left-wing baggage. What seemed to appeal to Groves, apart from the Oppenheimer charisma, was his apparent modesty about the work. He told the general that no one knew much and that they would have to make it up as they went along. Here was someone who would do what he was told to do. Groves was later asked to testify in the Oppenheimer hearings on behalf of Oppenheimer. He took the occasion to make very clear what he thought Oppenheimer's role was. He said,

"Dr. Oppenheimer was used by me as my adviser on that [different options for pursuing the work], not to tell me what to do, but to confirm my opinion. I think it is important for an understanding of the situation as it existed during the war to realize that when I made scientific decisions—in case there are any questions that come in on that—that outside of not knowing all the theories of nuclear physics, which I did not, nobody else knew anything either. They had lots of theories, but they didn't know anything. We didn't know whether plutonium was a gas, solid, or electric [sic— probably he said "a liquid"]. We didn't even know that plutonium existed, although Seaborg, I believe it was, claimed to have seen evidences of it in the cyclotron." Actually, in February 1941, Glenn Seaborg, McMillan, and their colleagues had produced plutonium by bombarding uranium with deuterons. By 1942 they had studied enough in trace amounts to understand some of its properties. Groves goes on,

"We didn't know what any of the constants that were so vital were. We didn't know whether it could be made to explode. We didn't know what the reproductive factor was for plutonium or uranium 235 [the number of neutrons emitted when these elements fissioned]. We were groping entirely in the dark. That is the reason that General Nichols [Kenneth D. Nichols was another army engineer who worked with Groves on the project] and myself were able, I think, to make intelligent scientific decisions, because we knew just as much as everybody else. We came up through kindergarten with them. While they could put elaborate equations on the board, which we might not be able to follow in their entirety, we knew just about as much as they did. So, when I say that we were responsible for the scientific decisions, I am not saying that we were extremely able nuclear physicists, because actually we were not. We were what might be called 'thoroughly practical nuclear physicists.'"

And then he adds,

"As a result of this experience, maybe because Dr. Oppenheimer agreed with me and particularly because of other questions that were raised, I came to depend on him tremendously for scientific advice on the rest of the project, although I made no effort to break down my compartmentalization. As you know, compartmentalization of information was my chief guard against information passing. It was something that I insisted on to the limit of my capacity. It was something that everybody was trying to break down within the project. I did not bring Dr. Oppenheimer into the whole project, but that was not only because of security of information—not him in particular, but all the other scientific leaders, men like Lawrence and Compton [Arthur Compton was a Nobel Prize–winning physicist who, with Conant and Bush, was one of the scientific leaders of the project] were treated the same way—but it was also done because if I brought them into the whole project, they would never do their own job. There was just too much of scientific interest, and they would just be frittering from one thing to another."

Oppenheimer was present in the room while General Groves was explaining his role. What he thought of Groves's description is not recorded. In any event, immediately upon being selected as director of the yet nonexistent laboratory, he adapted himself to the role. He at once agreed with Groves's notion that it should be a military facility with everyone there, including himself, in uniform. He seems to have gone so far as to have taken a physical examination for his entrance into the army and to have ordered uniforms. He was talked out of what would have been a folly by wiser heads like Rabi. It is hard to imagine, for example, Feynman as a corporal. What rank would Bohr have had? Would they have had to salute each other? In the event, the laboratory was formally administered by the University of California. There was a military presence on the site, but the scientists remained civilians.

It is important to emphasize that when Groves appointed Oppenheimer he had seen the FBI reports on him. He knew that Oppenheimer's brother Frank had been a member of the Communist party, as had Frank's wife. He knew that Oppenheimer's own wife, as well as his former lover Jean Tatlock, had been members. He knew that some of Oppenheimer's students had been members and that he had contributed substantial sums of money to Communist front organizations. He did not know the full details of Oppenheimer's relationship with Haakon Chevalier—one of the items that would prove to be most damaging in the hearing—but when he learned about them during the course of the war, this did not change his mind. On July 20, 1943, shortly after Los Alamos had gotten started, Groves wrote a letter to the War Department which said, "In accordance with my verbal directions of July 15, it is desired that clearance be issued for Julius Robert Oppenheimer without delay, irrespective of the information which you have concerning Mr. Oppenheimer. He is absolutely essential to the project." A great many people with derogatory information about them were "absolutely essential to the project." Serber once remarked that, if normal procedures had been applied, three-quarters of the people at Los Alamos, including very likely himself, would not have been cleared. My sense is that if the devil himself had turned out to have unexpected abilities in nuclear physics, Groves would have cleared him, but put him under constant surveillance.

So much has been written about wartime Los Alamos that it has become something of a cliché. But here are a few facts that have struck me over the years. The first is the duration of the enterprise. I find that when I tell people how long it lasted, they are surprised. The consequences were of such monumental importance that they are sure it must have gone on a lot longer than the twenty-seven months that were the actual time span. The project at Los Alamos began in the spring of 1943 and was over in the fall of 1945. The average age of the technical staff was twenty-nine. Oppenheimer was thirty-nine when it started. Not only were these young people making up the science as they went along, they also had to figure out what to do about the consequences of their work. In the beginning they were the only ones who understood the dimensions of the new force they had created. This was especially true after the successful test on July 16, 1945, at Alamogordo, in the New Mexican scrub desert. Rabi, who was older and more worldly than most of them, said that at first he was overwhelmed by the spectacle of the explosion, but then he got gooseflesh when he realized what this meant for humanity.

When Los Alamos began, Oppenheimer thought that a small enlargement of the group that met in Berkeley in the summer of 1942 would be enough to do the job. When it ended there were some three thousand people on the mesa. Never before, or since, has such a collection of scientific talent been assembled to carry out one task. I made a list of the people there who either had the Nobel Prize before they came or received it after they left. I am not sure my list is complete. Fermi and Bohr had Nobel Prizes when they came. Bethe and Rabi got theirs later. Rabi was not on the staff at Los Alamos, but he acted, on his frequent visits, as a sounding board for Oppenheimer. He told me that he did not go there full time because, as he put it, he was serious about winning the war. With the bomb one might win the war, but without radar one would lose it. Rabi chose to spend most of his time working on radar. The experimental physicists Norman Ramsey and Emilio Segré won the prize, as did the theorists Felix Bloch, Richard Feynman, and Bohr's son Aage, as well as the experimentalists Edwin McMillan, Owen Chamberlin, Fred Reines, and Luis Alvarez. Two of the more interesting cases were Val Fitch and Joseph

Rotblat. Fitch, who had been born on a cattle ranch in Nebraska, was sent to Los Alamos as a soldier. He had not finished college, but, dressed in army fatigues, he worked in one of the laboratories. He made such a good impression that he was urged to continue with his studies in physics. He was a professor at Princeton in 1980 when he won his Nobel Prize for Physics for his work on elementary particles. Joseph Rotblat, on the other hand, was born in Warsaw and emigrated to Britain when the war began. He came as a physicist with the British delegation that included Frisch and Peierls. In late 1944 he decided to leave. He had learned that the reconnaissance mission—Alsos—that Groves had sent to learn about the German nuclear weapons program had found that there was no such program. Rotblat had come to Los Alamos to defeat the Germans, and having decided that his mission had been accomplished, he left. After the war he devoted himself to the cause of making a nuclear world safer. In 1995 he won the Nobel Peace Prize.

Compared to what these scientists were used to, conditions on the mesa were primitive. There were five bathtubs for the whole community, located in the old school. The wartime apartment constructions had only showers. Water was always a problem. Sometimes when people turned on the taps, worms came out, and when it rained there was a sea of mud. It was like a sort of campout.* The one quality that runs through all the scientists' narratives

*Recently I made a trip to Los Alamos. Nothing looked familiar. In 1957 it was still a closed town surrounded by a fence. Now it is an open suburb of Sante Fe. I could not even find the building where I worked, to say nothing of any of the wartime structures. I also tried to find Oppenheimer's ranch in the Pecos. We apparently came within a few miles of it, but the road leading to the ranch is obscure and unpaved. How they got automobiles up it in the 1920s is a puzzle. The country is wonderfully beautiful—heavily forested—and bears little resemblance to the desert around Sante Fe. One understands what attracted Oppenheimer to it.

is enthusiasm. It may seem callous to say, but as hard as they worked building the bomb, they also had a great deal of fun. There were parties in which lab alcohol, 200 proof, was used to spike the punch. At one point Groves complained to Oppenheimer that people were having too many babies and overloading the medical facilities. Oppenheimer was in no position to do anything about that, especially since his second child, Katherine ("Toni"), was born there. She did not have a very happy life, and in a deep depression she committed suicide in 1977. For the people who spent those two years on the mesa, it was an experience that marked them for life.

A good deal of this experience had to do with Oppenheimer. It was as if all his gifts had been hoarded for this occasion. The all but instantaneous ability to comprehend and synthesize scientific ideas, which had terrorized his students because they could not keep up with him, was now channeled to making the project work. Hans Bethe, who is one of the foremost theoretical physicists of this era, told me that Oppenheimer's intellectual superiority was evident. He understood every aspect of the program from physics to the shop, and he could keep it all in his head. He also set the social tone. Pictures of the Los Alamos parties show him surrounded by attentive people. Even Edward Teller, who did not much like Oppenheimer—his wife actively disliked him—had to admit that he was an extraordinary laboratory director. During the course of his testimony at the 1954 hearing, testimony that was extremely damaging to Oppenheimer, Teller made this evaluation: "I would like to say that I consider Dr. Oppenheimer's direction of the Los Alamos Laboratory a very outstanding achievement due mainly to the fact that with his very quick mind he found out very promptly what was going on in every part of the laboratory, made right judgments about things, supported work when it had to be supported, and also I think with his very remarkable insight in psychological matters made just a most wonderful and excellent director." In offering this encomium, Teller seems to have forgotten just what happened at Los Alamos.

In addition to dealing with everything else, Oppenheimer found himself forced to deal with Teller. The trouble began almost immediately. One of the first things Oppenheimer did when he became director of the laboratory was to create divisions. One was the Theoretical Division, and he put Bethe in charge of it. Teller never got over this. I believe many of his later actions, including his testimony at the 1954 hearing, had this at their base. In his memoir, Teller writes, "When he told Bethe and me that he had named Hans to lead the division I was a little hurt. I had worked on the atom bomb project longer than Bethe. I had worked hard and fairly effectively on recruiting, and on helping Oppie organize the lab during the first chaotic weeks." But in the next paragraphs he makes it clear what a "little hurt" really meant. At one point he writes that Bethe came into his office to try to get him to work on something that was important to the laboratory, and Teller refused. He writes, "Fortunately for me, we never had to resume that conversation. Although Hans did not criticize me directly, I knew he was angry. Although I hoped that he would come to understand my position, he never did, and the incident marked the beginning of the end of our friendship. Oppenheimer got wind of our disagreement (whether because Hans told him or someone did, I do not know). He was much more sympathetic to what I was saying and apparently agreed that productive work requires a person to follow his strengths and inclinations. He told me to continue the variety of projects I was working on and encouraged me to look for new approaches and alternative ideas. He also assumed direct responsibility for determining the assignments for me and my group."

What I think happened is that Oppenheimer gave up on Teller as being a productive member of the Theoretical Division and simply decided to let him go off in a corner to pursue his super-bomb chimera. He managed to spare a little time each week to listen to Teller one-to-one. Rabi told me that one of the things he did when he came to Los Alamos to consult with Oppenheimer was to give him advice on how to handle the "Central Europeans."

Teller's refusal to do the work of the laboratory became particularly acute in the summer of 1944, when it seemed as if the whole project might collapse. This was the time when significant amounts of plutonium had started to arrive at Los Alamos from the reactors. To recapitulate how the plutonium is made in reactors:* a neutron from the fission of $U(235)$ is absorbed by a $U(238)$ nucleus to make $V(239)$. This nucleus is unstable and decays into neptunium—$Np(239)$—which in turn decays into plutonium—$Pu(239)$. This isotope of plutonium is reasonably stable. It has a lifetime of about forty thousand years. But if the plutonium is kept in the reactor too long, there is such a high flux of neutrons available that the $Pu(239)$ nuclei can absorb a second neutron and become $Pu(240)$.† This isotope is somewhat less stable, but the crucial thing is what it can decay into. Unlike $Pu(239)$, $Pu(240)$ can fission spontaneously—decay by fissioning into medium-

*Unlike uranium, there is no naturally occurring plutonium. It can't be mined. For a discussion of these matters, see, for example, Serber, 1992.

†What is known as "weapons' grade" plutonium is kept in the reactor only so long as less than 7 percent of the plutonium produced consists of the 240 isotope. "Reactor grade" plutonium, the kind you find in a power reactor, has, by definition, greater than 19 percent 240. Weapons of uncertain efficiency can be made using reactor-grade plutonium.

weight nuclei and neutrons, with considerable probability, without the fission being provoked by an external neutron. From the weapons point of view, this is catastrophic. It means that if you do not assemble the subcritical portions of a bomb using this reactor-produced mixture of Pu(239) and Pu(240) fast enough, spontaneous fission will start the chain reaction before the critical assembly is reached. The result will be an explosion of low yield—a "fizzle."

That there might be Pu(240) mixed in with the Pu(239), and that it might spontaneously fission, had been considered for some time, but it was only in April 1944 that the worst-case scenario was clearly realized. At first it was hoped that the experiments at the lab might be wrong, but on July 4 Oppenheimer announced, in a lab-wide colloquium, that the reactor plutonium was fatally contaminated with Pu(240). Until then it had been thought that the plutonium and uranium bombs could be developed along parallel tracks. Since the U(235) does not spontaneously fission at rates high enough to pose a problem, one could make use of a "gun" which shot the subcritical parts together at the speeds characteristic of artillery shells. While this had not been tested—indeed it never was tested before the Hiroshima bomb where, evidently, it worked—the technology seemed relatively straightforward. If one tried to do this with the reactor-created plutonium, the assembly was too slow. This left a small number of equally unpalatable alternatives. One could try to mix the plutonium with uranium, which would produce a bomb of relatively low yield. One could try to separate the plutonium isotopes by building another Oak Ridge. This would push the plutonium bomb off to the indefinite future. Or one could try a new technique which had only been speculated on in a marginal way to this point—implosion.

The basic idea of implosion is to take, say, a sphere of plutonium at normal density which is subcritical, then crush—implode—the sphere from all sides so that it becomes smaller and denser. This dense sphere can now go critical and begin the explosive chain reaction. The problems were immense. How to crush the sphere uniformly? If one can do it, can one do it fast enough so that predetonation does not take place? Indeed, how to start the detonation of the imploded sphere? In July 1944 no one knew the answers to any of these questions or even how to go about finding the answers. But Oppenheimer and his advisers decided that this was the only choice. Oppenheimer got Groves's permission to reorganize the laboratory completely. A small group would continue working on the uranium bomb, but everyone else would work on implosion—everyone but Teller. He had been an early champion of implosion, but when it became the work of the laboratory, he would not do it. This repeated itself with the hydrogen bomb after the Ulam-Teller invention was adopted by Los Alamos as the way to proceed. Teller walked away claiming that he did not like the organization of effort. But in both cases it worked anyway.

From the time of the reorganization in July 1944 to the time the implosion bomb was tested at Alamogordo was one year. Indeed, design of the implosion bomb was frozen in February 1945. This was the design that Klaus Fuchs was able to transmit to the Russians. It probably saved them a year or two; the equivalent of the Ulam-Teller idea they found on their own, with some help from Fuchs. Whole new technologies were invented during this period. To make the implosion uniform, a new concept of explosive lenses was invented—shaped charges that focused the detonation wave. New techniques were invented to measure the uniformity of the implosion. The first attempts were terrible—looking more like an automobile wreck than spherical compression. It was decided that instead of a hollow sphere, a solid sphere would be easier to implode uniformly. This was called the Christy gadget, named after Robert Christy, one of Oppenheimer's Berkeley students who designed it. It was a Christy gadget that was dropped on Nagasaki. During all of this effort, Oppenheimer worked himself beyond what anyone would have thought possible. Groves was worried that he might work himself sick. At the end he weighed 115 pounds. But Groves underestimated Oppenheimer's toughness.

By December 1944 a base camp had been set up in Alamogordo for the test of the first plutonium implosion bomb. Oppenheimer called the test "Trinity." In 1962, General Groves asked him why, and Oppenheimer wrote back, "I did suggest it. . . . Why I chose the name is not clear, but I know what thoughts were in my mind. There is a poem of John Donne, written just before his death, which I know and love. From it a quotation:

> . . . As West and East
> In all flat maps—and I am one—are one.
> So death doth touch the Resurrection.

That still does not make Trinity, but in another, better known devotional poem Donne opens, 'Batter my heart, three person'd God.' Beyond this, I have no clues whatever."

The debate over whether the bomb should have been built, or whether once built it should have been used on a city or just demonstrated, will never be resolved. After the Germans were defeated, Robert Wilson, who had been a colleague of Oppenheimer's at Berkeley and who was one of the more thoughtful physicists, organized a meeting to discuss whether they should go on. Oppenheimer, for reasons Wilson did not understand, tried to persuade him not to have the meeting. But he did, and some fifty people came, including Oppenheimer. He argued that the United Nations was in the process of being organized and that the world should have an understanding of this new force, which it could have only if the bomb were built and demonstrated. Wilson and the others were persuaded and continued to work. Rabi and Oppenheimer discussed the feasibility of a demonstration and decided that, in the middle of a war, it was unworkable. Who would come? What would it mean? What would happen if the test were a failure? The technology was still very new. After the bombs were dropped, most of the people who helped build them shared Bethe's view. First, he had a sense of satisfaction that something they had worked on had helped to win the war. Next he thought, "What have we done? What have we done?" Finally, he was determined to do everything he could to keep it from happening again.

October 16, 1945, was Oppenheimer's last day as director. The laboratory had begun to disband, but on November 2 he gave a talk to some five hundred of the remaining people on the site — sponsored by the Association of Los Alamos Scientists. In it he presented his hopes and fears for what was to come. At one point he said,

"In considering what the situation of science is, it may be helpful to think a little of what people said and felt of their motives in coming into this job. One always has to worry that what people say of their motives is not adequate. Many people said different things, and most of them, I think, had some validity. There was in the first place the great concern that our enemy might develop these weapons before we did, and the feeling — at least, in the early days, the very strong feeling — that without atomic weapons it might be very difficult, it might be impossible, it might be an incredibly long thing to win the war. These things wore off a little as it became clear that the war would be won in any case. Some people, I think, were motivated by curiosity, and rightly so; and some by a sense of adventure, and rightly so. Others had more political arguments and said, 'Well, we know that atomic weapons are in principle possible, and it is not right that the threat of their unrealized possibility should hang over the world. It is right that the world should know what can be done in their field and deal with it.' And the people added to that that it was a time when all over the world men would be particularly ripe and open for dealing with this problem because of the immediacy of the evils of war, because of the universal cry from everyone that one could not go through this thing again, even a war without atomic bombs. And there was finally, and I think rightly, the feeling that there was probably no place in the world where the development of atomic weapons would have a better chance of leading to a reasonable solution, and a smaller chance of leading to a disaster, than within the United States. I believe all these things that people said are true, and I think I said them all myself at one time or another."

Then he added, "But when you come right down to it, the reason that we did this job is because it was an organic necessity. If you are a scientist you cannot stop such a thing. If you are a scientist you believe that it is good to find out how the world works; that it is good to turn over to mankind at large the greatest possible power to control the world and to deal with it according to its lights and its values."

And dealing with it is what we have been trying to do ever since.

Just after the Japanese surrendered, the Oppenheimers went to their ranch in the Pecos. He had an exchange of letters with Haakon Chevalier, whom he had not been in contact with. In his response to Chevalier's letter he wrote,

"Dear Haakon,

"Your letter, your marvelous warm letter, was one of the very few things that brought warmth to us over these troubled days. A

89

few days after the surrender we went over to our ranch for a few days' time of solitude, horses, and the slow return to sanity. We are not sure we will be coming back to Berkeley for permanent despite the ties that make us want to. We are not too sure of anything personal, longing both of us for stability, yet knowing we have been put in a time and a place where we may not be able, in conscience, to attain it. The circumstances are heavy with misgiving and far, far more difficult than they should be had we the power to remake the world to be as we think it."

For Oppenheimer the world was already remade. He would now become a public man—his next adaptation.

miliar with it. The notes read, "Oppenheimer remarked that it was not safe to assume that everybody was familiar with this, but it was also not safe to assume that this is any reason for discussing it." The other thing that struck me was an exchange with Edward Teller. There was an apparent contradiction between two sets of experimental results, and the notes say, "Oppenheimer asked Teller if he would not elaborate upon his opinion of this matter, and Teller proceeded to do so." This was in January 1954. The hearing before the Personnel Security Board of the Atomic Energy Commission, which was held in Washington, began on April 12, 1954, fewer than three months later. I am sure that when Oppenheimer asked Teller to "elaborate upon his opinion," he had no idea of Teller's role in what would become a successful attempt to destroy Oppenheimer's reputation.

Oppenheimer's attention could not have been entirely on the physics at the conference. In a multipage letter dated December 23, 1953, from Major General K. D. Nichols, the general manager of the Atomic Energy Commission, Oppenheimer was informed that he was now regarded as a security risk. The letter, which outlines the charges, begins:

"DEAR DR. OPPENHEIMER: Section 10 of the Atomic Energy Act of 1946 places upon the Atomic Energy Commission the responsibility for assuring that individuals are employed by the Commission only when such employment will not endanger the common defense and security. In addition, Executive Order 10450 of April 27, 1953, requires the suspension of employment of any individual where there exists information indicating that his employment may not be consistent with the interests of national security.

"As a result of additional investigation as to your character, associations, and loyalty, and review of your personnel security file in light of the requirements of the Atomic Energy Act and the requirements of Executive Order 10450, there has developed considerable question whether your continued employment on Atomic Energy Commission work will endanger the common defense and security and whether such continued employment is clearly consistent with the interests of the national security. This letter is to advise you of the steps which you may take to assist in the resolution of this question. . . ."

The timing of this letter, a few weeks before the Rochester conference, raises the obvious question of what effect it had on Oppenheimer at the meeting. In particular, did he discuss it with people there? One finds no hint in the conference transcripts themselves. Oppenheimer's remarks are devoted only to physics. But I was able to ask Bethe, who was there. His memories were vivid. He recalled Oppenheimer's distress. He told Bethe that in this confrontation he was convinced he would lose. He was under no illusions.

Long after the hearing, someone remarked to Oppenheimer what a tragedy the hearings had been. Oppenheimer replied that they weren't a tragedy but rather a farce. Possibly the hearings themselves, with their Byzantine, interminable questions about what seemed to be trivia, might be thought of as farcical, but the events leading up to them were tragic. Oppenheimer seemed to sow the seeds of his own destruction. While Los Alamos brought out all the best sides of his complex personality, these events brought out the worst. To understand this we must return to what happened after Hiroshima and Nagasaki.

The first feeling of most of the physicists at Los Alamos upon hearing of Hiroshima over a loudspeaker in the laboratory was that of relief. The device had worked. It was not obvious that it would; it was a uranium bomb, which had never been tested. The next feeling was that of horror. As Bethe said, "What have we done?

What have we done?" Robert Wilson, who had organized a meeting after the Germans were defeated to discuss whether the laboratory should go on making the bomb, became physically sick. When Oppenheimer learned of this he knew that the reaction to having made the bomb had begun, at least at Los Alamos. But Los Alamos was only one part of what was called the Manhattan Project. Other parts included Oak Ridge, where the uranium was being separated; Hanford, Washington, where the reactors were making plutonium; and a smaller project at the University of Chicago—the Metallurgical Laboratory, which was studying the metallic properties of uranium and plutonium.

In charge of the chemistry division of the Metallurgical Laboratory was a man named James Franck. He had won the Nobel Prize for Physics in 1925 and had come to the United States in the 1930s as a refugee from Germany. Even before the 1945 test at Alamogordo, Franck and his colleagues in Chicago were already concerned about what nuclear weapons would mean. This resulted, in early June 1945, in the drafting of a document which came to be called the Franck Report. The premise of the report, and indeed of the thinking about these matters by the scientists who worked on the bomb, was that it could not be kept secret. This did not have to do with espionage but with the understanding that there was nothing in the technology that could not be rediscovered elsewhere. Nonetheless there was general surprise when the Russians tested their first nuclear weapon—which we called "Joe 1" in honor of Stalin—on August 29, 1949. It was not announced by the Russians, but it was detected on September 3 by a B-29 bomber acting in the guise of a weather plane. The bomber picked up samples of the radioactive fallout, which was then analyzed by scientists.

In any event, the arrival of the Russian bomb came faster than

most scientists had expected. The speed was in part due to successful espionage, but it should also be mentioned that when the Soviets occupied parts of Germany they sought out the nuclear
scientists and shipped them east. While it is true that the Germans
had not gotten anywhere with designing a weapon, they had made
progress in things like isotope separation, which was useful to the
Russians. President Truman's reaction to the news of the Soviet
test was interesting. He simply did not believe that "those Asiatics"
could build a bomb. Before he announced it to the country on
September 23, it seems that he insisted that several of the scientists
who had made this claim sign a document stating that they really
believed the Russians were capable of such an achievement. But,
as I have noted, there was no doubt in the minds of the people who
helped draft the Franck Report in 1945, that it would be only a
matter of time.

The Franck Report took the position that a first use of nuclear
weapons by the United States would be a mistake. The report
reads: "If the United States would be the first to release this new
means of indiscriminate destruction upon mankind, she would
sacrifice public support throughout the world, precipitate the race
of armaments, and prejudice the possibility of reaching international agreement on the future control of such weapons. Much
more favorable conditions for the eventual achievement of such an
agreement could be created if nuclear bombs were first revealed to
the world by a demonstration in an appropriately selected unin-
habited area. . . ."

As it happened, the writers of the report had a significant platform on which to place it. On April 25, at the same time plans for
the United Nations were being formulated in San Francisco, Secretary of War Henry Stimson and General Groves briefed President Truman, for the first time, about the atomic bomb—a project

of which he had known nothing. At Stimson's suggestion, Truman authorized the creation of what was known as the Interim Committee to advise him on all aspects of the new weapon. This was a small committee which included people like Vannevar Bush and James Byrnes. But they appointed a scientific advisory panel that included Oppenheimer and Fermi. This panel was given the Franck Report to consider. It rejected its proposals. In particular it endorsed the use of a nuclear weapon on Japan. Precisely where Oppenheimer stood was never made clear, at least by him. It is clear, however, that he did not strongly object—if at all—to the military use of the bomb. In fact he participated in the target committee that selected the Japanese cities.

This is not to say that Oppenheimer was indifferent to the disposition of nuclear weapons. Quite the contrary, he gave it a great deal of thought, and in this Niels Bohr was his mentor. Bohr had come to Los Alamos at the very end of 1943. He had escaped from Denmark, via Sweden, a couple of months earlier and had been flown to England in an RAF fighter. Before the summer of 1943 he had been certain that nuclear weapons were a practical impossibility. In February of that year he had declined an invitation, tendered to him via the underground, to come to England and work on the project, because he thought the prospects of success were so poor. But in the summer of 1943 he was visited by a German physicist named Hans Jensen who had connections with the German nuclear physicists. Jensen brought with him either a drawing, or enough information for Bohr to make a drawing, of a device that both he and then Bohr thought was a design for a nuclear weapon.

When the much alarmed Bohr came to the United States, he had this drawing with him. He showed it to General Groves. Now it was Groves's turn to become alarmed. He ordered an immediate full-scale analysis of the drawing, which took place in Los Alamos on December 31, 1943, at a meeting at which both Bohr and his son Aage were present. Bethe and Teller took one look at the drawing and saw at once that it was the drawing of a reactor—one, it turned out, that Heisenberg had designed but was never able to produce a self-sustaining chain reaction with—and was in no way any kind of nuclear weapon. They wrote a report which Oppenheimer sent to Groves the next day.

Bohr was then given a crash course in what had been happening at Los Alamos. Oppenheimer assigned Feynman as his tutor. Bohr, in turn, explained his ideas to Oppenheimer for a nuclear future based on an open world with a free exchange of information. The following May he was able to explain his vision to Winston Churchill. The meeting, which was a disaster, inspired Churchill to contemplate having Bohr locked up. He had more luck with Roosevelt, at least a more cordial reception. But the president died before Bohr had a chance to talk with him again. Meanwhile Oppenheimer had been discussing the nuclear future with a few people, especially Rabi.

The most important discussions took place just after the war. When I interviewed Rabi in the early 1970s for my *New Yorker* profile, they were still fresh in his mind. Many of our interviews took place in Rabi's study, the very room in which he and Oppenheimer had had some of their talks. Rabi recalled, "Oppenheimer and I met frequently and discussed these questions thoroughly. I remember one meeting with him, on Christmas day of 1945, in my apartment. From the window of my study we could watch blocks of ice floating past on the Hudson. We were then developing the ideas that became the basis of the Acheson-Lilienthal report."

In January 1946 the undersecretary of state, Dean Acheson, had been asked to chair a committee that would recommend the atomic-energy policy which the United States would propose in the recently formed United Nations. The committee consisted of people like James Conant and General Groves. They decided they needed consultants with various kinds of technical expertise, and a board was created with David Lilienthal as its chairman. Lilienthal was then chairman of the Tennessee Valley Authority, TVA. He was a lawyer who at the age of thirty-three, in 1933, had been appointed one of the original directors of the TVA, which became the model for public power projects in the United States. During the war he helped to locate the site in Tennessee on which Oak Ridge was built. He asked Oppenheimer to become a member of this advisory board. As Lilienthal's journals make clear, he simply hero-worshiped Oppenheimer. He had never met anyone quite like him. Not long before his death in 1981, I met Lilienthal. We talked about Oppenheimer, and it was clear that he still felt the same way.

The Acheson-Lilienthal report, which bears all the hallmarks of Oppenheimer's style, was issued publicly on March 28, 1946. Rabi described its intentions: "It was the most extraordinary kind of liberal position for the United States to take—that there should be no private and no national ownership of uranium and other fissionable materials. They should all be internationally owned under United Nations supervision. This was the actual American position, at a time when we had a monopoly on the atomic bomb." But, the person whom President Truman chose to present the plan, or some version of it, in the United Nations was neither Acheson nor Lilienthal but the seventy-five-year-old financier Bernard Baruch. When Acheson, Lilienthal, and Oppenheimer heard of this, they were horrified. Not only did it seem to them that Baruch did not have the vigor to promote such a program, but it was not clear to them that he even accepted its basic tenets. As events soon showed, he didn't. Oppenheimer was so distressed that Acheson decided to take him to meet Truman, possibly to persuade the president to replace Baruch. This meeting was a disaster, comparable only to Bohr's meeting with Churchill. Acheson later described it in his memoir, *Present at the Creation*. Of Oppenheimer he writes, "Hesitant and cheerless, he seemed so different from his reputation that Truman wanted to know what was the matter. 'I feel we have blood on our hands,' Oppenheimer mumbled. 'Never mind,' said Truman, 'it'll all come out in the wash.'" Then Truman went on, "When will the Russians be able to build the bomb?" "I don't know," Oppenheimer replied. Truman then answered his own question by saying "Never." Afterward he told Acheson, "Don't you bring that fellow around again. After all, all he did was make the bomb. I'm the guy who fired it off."

On June 14, 1946, Baruch presented to the United Nations what was essentially his plan. It differed dramatically from the Acheson-Lilienthal proposal and was formulated in such a way—this may have been the intention—that the Russians could not possibly accept it. For example, it called for total disarmament and the abrogation of the veto power on matters of atomic energy. There would be no international depository for the ingredients of nuclear weapons. This ended Oppenheimer's formal attempts to influence government policy on the international control of nuclear weapons.

Rabi told me about a conversation he had with a Russian physicist by the name of Dimitri Skobeltsyn, who came to Columbia to talk informally with him about what was happening in the United Nations. "Skobeltsyn," Rabi said, "was a good physicist and an intelligent man. He would come to me with a copy of some fire-eating speech by some senator. I would try to tell him not to pay any attention to it. But how could he believe me? They thought it was some capitalist trick and that we wouldn't adhere to the agreements. They felt that it was all a way of stopping them, and this belief was probably reinforced by the rather belligerent way Baruch actually presented the proposal. Or I think they felt that, and certainly it was natural for them to feel that."

In retrospect it is clear that the Russians would not have accepted any proposal for internationalizing nuclear energy after the bombing of Hiroshima. Stalin was determined that the Soviet Union would have an atomic bomb at any cost. Even Russians like Andrei Sakharov, who later turned against the Soviet nuclear weapons program, at the time felt it was necessary for the survival of the Soviet Union. One must also recall that not long before Baruch spoke at the United Nations, Churchill had spoken at Westminster College in Fulton, Missouri, with Truman on the platform, and had declared that the Soviet Union had now withdrawn behind an "Iron Curtain." The cold war was on.

Meanwhile Oppenheimer was rapidly becoming a celebrity, almost like a movie star. It was a combination of his association with this great new power—the atomic bomb—his striking looks, and his unusual use of language. He had appeared on the cover of *Time* magazine and was the subject of a great variety of popular articles. What's more, he had become one of Philippe Halsman's "jumpers." Halsman was a noted photographer whose portraits of people like Einstein and Marilyn Monroe have become classics. At some point he got the notion that the way people jump reveals deep insights into their psyches. In his *Jump Book* he wrote, "In a jump the subject, in a sudden burst of energy, overcomes gravity. He cannot simultaneously control his expressions, his facial and his limb muscles. The mask falls. The real self becomes visible. One has only to snap it with the camera." In any event, by the time he was finished he had gotten some two hundred people to jump for him. They included Richard Nixon, the Duke and Duchess of Windsor, Peter Ustinov, and Carol Channing. Into this mix came Oppenheimer.

In studying the photograph of the jumping Oppenheimer, I am persuaded that Halsman may have been on to something. First of all, there are the clothes. Oppenheimer is wearing a three-piece suit. If one looks closely at the open jacket, one can see the Langrock's label. Oppenheimer had his suits tailored for him at Langrock's, Princeton's bespoken tailor. He also had a special capelike green loden coat. Mr. Decker, who was Langrock, had in his store an autographed picture of Oppenheimer wearing one of his suits which he would show you if he learned you were at the Institute. I doubt that Mr. Decker had thought to design his suits for jumping. In the photograph, Oppenheimer is looking heavenward with a finger extending in that general direction. The fingers on the other hand are pointing down. He does not seem to have levitated very far off of whatever surface he is jumping from. Halsman commented, ". . . In the Princeton Institute for Advanced Studies [sic], Dr. J. Robert Oppenheimer jumped for me, the arm outstretched and the hand extended toward the ceiling. 'What do you read in my jump?' he asked. Jumpology had a simple explanation for it, but I felt puzzled. I looked at Dr. Oppenheimer's Spartan study, thought of his life dedicated to science and decided that the rule was not applicable." Halsman's rule was that an outstretched arm was a sign of ambition. "Probably there was another motivation. 'Your hand pointed upward,' I hazarded, 'maybe you were trying to show a new direction, a new objective.' 'No,' said Oppenheimer, laughing, 'I was simply reaching.'"

Some of the people who had known him before he became a celebrity did not like the new Oppenheimer. For example, Philip Morrison, who had been one of Oppenheimer's students at Berkeley and then had gone on to Los Alamos, reported on a conversation with the new Oppenheimer. It was filled with references to "George" and "Dean"—"George" this and "Dean" that. It was only a bit into the conversation that Morrison was given to understand that "George" was General George C. Marshall, and "Dean" was Dean Acheson. While this sort of condescension might not have caused any irreparable damage if Oppenheimer had limited it to former students, it became part of his mannerism when he dealt with figures in the government, and there it did cause irreparable damage. A very important instance involved Lewis Strauss.

Strauss, who pronounced his name *straws*, insisted on being called "admiral"—in 1944 he had been named a rear admiral in the naval reserve. He had been born in Charleston, West Virginia, in 1896. His father had a wholesale shoe business. When Strauss was seventeen, his father tapped him to sell shoes to shoe stores. His job was to open up new territories. He had managed to save up enough money to go to college when the First World War intervened. He had wanted to study physics. At that time Herbert Hoover had been appointed by President Wilson to manage supplies, and Strauss volunteered his services. Hoover was impressed by the fact that Strauss did not ask for a salary. Very shortly, Hoover made him his personal assistant.

After the war Strauss joined the financial firm of Kuhn, Loeb and over the next decades became a millionaire. In 1925 he was commissioned in the naval reserve as an intelligence officer, and in 1941 was called to active duty. Eventually he became a special assistant to the secretary of the navy. This would lead to Strauss's involvement with nuclear energy when the services began to consider how to employ the bomb. On the surface he was a man with gracious manners, but he had a gigantic ego and was unwilling to accept opposing points of view. Soon he and Oppenheimer were at loggerheads.

After the war there was a struggle over whether the nuclear program would remain in military hands or become a civilian enterprise. Oppenheimer's position on this was, at first, perhaps deliberately ambiguous. He seemed quite prepared to accept military control if this was the decision; but most of the scientists who had worked on the bomb were absolutely determined to take it away from the military, so he changed his mind. Pressure from the scientists resulted in a bill sponsored by the freshman Connecticut senator Brien McMahon, which he presented at the end of December 1945. It became law the following April—the Atomic Energy Act of 1946—but its provisions went into effect only on January 1, 1947. This remarkable piece of legislation created an Atomic Energy Commission, the AEC, which had jurisdiction over all aspects of the program. One of the paragraphs of the act reads,

"The Commission is authorized and directed to have custody of all assembled or unassembled atomic bombs, bomb parts, or other military weapons presently or hereafter produced, except that upon the express finding of the President that such action is required in the interests of national defense, the Commission shall deliver such quantities of weapons to the armed forces as the President may specify." Keep in mind that this was a *civilian* agency.

The act also specified that the commission would be composed of five members with a chairman selected by the president. All the commissioners would be designated by the president with the advice and consent of the Senate. The president could determine their length of service. Truman chose Lilienthal as chairman and Strauss as one of the commissioners. Different reasons have been given for Strauss's selection. It is said that Truman wanted to appoint a Republican to balance the panel. Strauss had also served on an interservice committee that had considered nuclear weapons.

During the course of his service as commissioner, Strauss made the oft-quoted prediction that, because of nuclear energy, electricity would become too cheap to meter. Soon after the act went into effect, the commission created a nine-member group of advisers—the General Advisory Committee, GAC—and Oppenheimer was selected to it. So were Conant, Rabi, and Fermi, among others. The group elected Oppenheimer chairman. It was in this capacity that Oppenheimer had his first major confrontation with Strauss. The issue appears relatively banal.

In 1947 some of the Manhattan Project scientists thought it might be a useful gesture toward international goodwill if the United States offered small amounts of radioactive isotopes to friendly countries to be used in research. These isotopes could be made in quantity only in reactors, and at the time the United States had the monopoly on reactors. No one could think of any military uses for such isotopes—no one, that is, except for Strauss. Since the law now required approval by the Atomic Energy Commission for distributing these isotopes, it came up for a vote. The commissioners voted four to one—Strauss the lone dissenter—for the transfer. Naively, one might think, this would have ended the matter—but one was dealing with Strauss. He waited two years. In June 1949 the Norwegian Royal Defense Research Establishment wanted a small amount of the isotope Iron(59) to use in monitoring steel manufacture. Strauss learned that one of the putative Norwegian research workers had possible Communist ties. This gave him the leverage needed to bring the matter up with the conservative Republican senator Bourke Hickenlooper of Iowa, the chairman of the Joint Committee on Atomic Energy. On June 13,

Hickenlooper held a public Senate hearing. All the AEC commissioners were there, including of course Strauss. The hearing was filmed.

Oppenheimer, as chairman of the AEC Advisory Committee, testified in favor of the isotope export. If he had been content simply to make his case, the appearance might have been uneventful. But he chose to make a fool of Strauss. This was done by indirection, but the intent was clear. When asked, for example, if the isotopes could be used for military applications of atomic energy he replied, "No one can force me to say that you cannot use these isotopes for atomic energy. You can use a shovel for atomic energy; in fact you do. You can use a bottle of beer for atomic energy. In fact you do. But to get some perspective, the fact is that during the war these materials have played no significant part, and in my knowledge, no part at all." Later he commented that they were much less important for military applications than electronic devices, but much more important than vitamins.

It was not only what he said but the way he said it. I was able to study Oppenheimer's behavior for two years in our Institute physics seminars. He sat in the front row, and if he made what he thought was a witty comment he would look around to make sure that we had all taken it in. Many of these comments had a sarcastic edge. (He once got me. The topic was neutron stars and, for no reason that makes any sense to me now, I asked the speaker if there might be mu-mesons in the cores of neutron stars. I think I was then working on mu-mesons.* Oppenheimer fielded the question by saying sarcastically, "Mu-mesons are found in Berkeley and not

*I am employing the terminology of the time. We would now call them "muons" or "mu leptons."

in the stars." Incidentally, if present ideas are right, there are a lot more exotic particles to be found in the cores of neutron stars than mu-mesons.) In any event, Oppenheimer's performance at the hearing was witnessed by the AEC's general counsel, Joseph Volpe. He had plenty of opportunity to study Strauss while this was going on. He reported that he had never seen a clearer expression of hatred on anyone's face. Strauss was livid. When he had finished testifying, Oppenheimer came over to Volpe and asked, "Well, Joe, how did I do?" To which Volpe replied, "*Too* well, Robert. Much too well." Oppenheimer had made a dangerous enemy, as the next several years would reveal.

When the AEC took over the Manhattan Project on January 1, 1947, there was the residual question of clearances. The people who had worked in the Manhattan Project had been cleared, sometimes reluctantly, often because of wartime exigencies. What should their status be now? In his testimony in the 1954 Oppenheimer hearing, David Lilienthal, who was the AEC chairman in 1947, described what happened. He noted that there was a kind of grandfather clause in the Atomic Energy Act which allowed these people to preserve their clearances pending reexamination on a case-by-case basis. Lilienthal remembered that on Saturday, March 8, he received a call from J. Edgar Hoover informing him that he was sending over the FBI files of Oppenheimer and his brother.

Recall that, in 1936, Frank Oppenheimer and his new wife Jacquenette Quann—"Jackie"—had, against Oppenheimer's advice, joined the Communist party. Oppenheimer wasn't very enthusiastic about the marriage either. Frank Oppenheimer had been cleared by Groves to work at Los Alamos, relying on his brother's assertion that Frank was no longer a party member.

Hoover was adamant that Frank Oppenheimer's clearance should not be extended—and this is what happened. In June 1949 he and his wife appeared before a closed session of the House Un-American Activities Committee. They admitted that they had been Communists before the war but had long since resigned from the party. Frank's testimony was leaked to the press, and he lost his job at the University of Minnesota because he had previously lied to them about being a party member. He was not able to find another academic job. Using part of the proceeds from the sale of the van Gogh *First Steps (After Millet)* he had inherited from his father, he bought a piece of land in Colorado on which he raised cattle. He taught for a while at the University of Colorado in Boulder. But in the mid-1960s he got the idea of a novel kind of science museum in which one could actually interact with the exhibitions. In 1969 he founded the Exploratorium in San Francisco, which became a model for participatory science museums, and was its director until his death in 1985. It cannot have been easy to have been Robert Oppenheimer's younger brother.

As for Oppenheimer himself, the extent of the files was something of a shock to Lilienthal. He knew that Oppenheimer had had a "left-wandering" bohemian past, but he had no idea of the magnitude. He realized that this was something too important to do on his own, so the following Monday he called all the commissioners, including Strauss, together for a meeting. He notes, "It was very informal. We had this file which I requested all the commissioners to read. It was not necessary to request them because it was obviously a matter of great interest and importance. Instead of delegating this to someone else, it seemed clear that we should do the evaluating, since the responsibility of deciding what should be done, if anything, was ours. So we did begin a reading of the file around the table in my office in the New State Building, and later as time went on, members would take all or parts of their file to their offices and so on."

They noticed that while the file contained all the derogatory information, it contained no references to what Oppenheimer had accomplished during the war. Lilienthal then solicited statements from General Groves, Secretary of War Robert Patterson, Conant, and Bush. They were uniform in expressing their confidence in Oppenheimer's loyalty and fitness for service. To make absolutely sure, Lilienthal contacted Hoover to see if there was additional derogatory information that would cause Hoover to recommend against clearance. Hoover said he was troubled only by the incident connected with Haakon Chevalier (more about this shortly) and the fact that Oppenheimer had not fully reported it in a timely fashion. Lilienthal remarks, "Beyond that there was no further comment about the file. So we left with no suggestion from Mr. Hoover that further investigation ought to be carried on or that the file was incomplete, that there were things we didn't know about." On this basis, clearance was granted to Oppenheimer with no dissent, including that of Strauss. The importance of this action is made clear in a colloquy between Lilienthal and one of Oppenheimer's lawyers, Samuel J. Silverman, at the 1954 hearing. In the dialogue, Silverman is "Q."

> Q. Let me interrupt you for a moment. You have seen the Commission's letter of December 23, 1953, which suspended Dr. Oppenheimer's clearance.
> A. I have.
> Q. So far as you can recall what is the relation between the derogatory information contained in that letter and the material that was before you sent to you by Mr. Hoover in 1947?
> A. From my careful reading of the Commission's letter and my

best recollection of the material in that file, and the charges cover substantially the same body of information—

Q. Except for the hydrogen bomb stuff, of course.

Let us be clear about what is being said here: after a careful examination of the file in 1947, Oppenheimer was cleared by the Atomic Energy commissioners. But in December 1953 his clearance was revoked on the basis of the same body of information, "except for the hydrogen bomb stuff." Our task is then twofold: we must explain what happened between March 1947 and December 1953 that brought this about, and we must explain the "hydrogen bomb stuff." We start with the latter.

We have seen that during the war Edward Teller and a small group of acolytes pursued the grail of the "classical super," a device in which a fission bomb was to be placed in a container containing fusible light isotopes such as deuterium. The problem with this, as we have seen, was that the light isotopes could not be maintained at a high enough temperature to ignite a substantial amount of the material. On the other hand, a program to ignite small amounts to "boost" a fission explosion was successful. The quest for the super persisted during the years shortly after the war, though Teller found it difficult to get many people to work on it. As Oppenheimer—and no doubt others—pointed out, one of its difficulties was that there was no way to test proposals for constructing it, experimentally, without actually igniting large amounts of the fusible material, which was the very thing one was trying to figure out how to do. There were no computers then available to simulate these conditions.

This was essentially the status of things in August 1949 when the Russians made their first successful test. For Teller and others, especially in the military, news of this test produced almost a state of panic. Despite the fact that the United States had not been able to construct a hydrogen bomb, scientists and officials assumed that the Russians might, and that this would leave this country in a highly vulnerable state. Teller called Oppenheimer, who told him to "keep his shirt on"—advice that Teller was not prepared to take. The fact that we then had an arsenal with some two hundred nuclear weapons of various sizes was not, for Teller and others, sufficient deterrence. All of this was a prelude to one of the pivotal moments in the history of the hydrogen bomb, a meeting of the AEC advisers—the GAC—that took place over two days in Washington beginning on October 29.

Before this meeting, Oppenheimer wrote a letter to his fellow adviser Conant, whom he addressed as "Uncle Jim." In part, here is what he says:

"We are exploring the possibilities for our talk with the President on October 30th. All members of the advisory committee will come to the meeting Saturday except Seaborg, who must be in Sweden, and whose general views we have in written form. [Glenn Seaborg did not attend the meeting. Much was made in the 1954 hearing of whether Oppenheimer made Seaborg's views sufficiently known to the other advisers. But, as Seaborg's own letter shows, he felt the other advisers might have a better grip on the problems than he did. There is no indication he would have voted against the majority.] Many of us will do some preliminary palavering on the 28th.

"There is one bit of background which I would like you to have before we meet. When we last spoke, you thought perhaps the reactor program offered the most decisive example of the need for policy clarification. [The "super" proponents had actively agitated to build a new reactor with a very high flux of neutrons. This would be used to make the super-heavy isotope of hydrogen-tritium, a fusible element.] I was inclined to think the super might also be relevant. On the technical side, as far as I can tell, the super is not very different from what it was when we first spoke of it more than 7 years ago; a weapon of unknown design, cost, deliverability* and military value. But a very great change has taken place in the climate of opinion. On the one hand, two experienced promoters have been at work; i.e. Ernest Lawrence and Edward Teller. The project has long been dear to Teller's heart; and Ernest has convinced himself that we must learn from Operation Joe [the Russian test] that the Russians will soon do the super, and that we had better beat them to it."

Before giving the conclusion of this letter, I want to comment on the question of "military value." When asked about the "super," Oppenheimer had often said that one of its problems was that the targets were too *small*. This may seem like a very strange observation until one considers the numbers. The bomb that destroyed Hiroshima had a blast power equivalent to 15,000 *tons* of TNT. In contrast, the first successful U.S. hydrogen bomb tested—Ivy Mike—which evaporated Elugelab Island on the Eniwetok Atoll on October 31, 1952,† had a blast power equivalent to about 10.5 *million* tons of TNT. This hydrogen bomb could devastate an area about a thousand times the area devastated by the Hiroshima bomb. But there aren't cities this large, unless you count very large metropolitan areas. There are certainly no cities of this size in the Soviet Union. If anything, some U.S. metropolitan areas might qualify. Thus one was proposing to design a weapon—if you want to call it that—that would have been more useful for an adversary than for ourselves.

---

*The transcribers have rendered this "deliberability," but it is clear what was meant.

†This is Greenwich Mean Time. Locally it was a day later.

Oppenheimer concludes his letter, "What concerns me is really not the technical problem. I am not sure the miserable thing will work, nor that it can be gotten to a target except by ox-cart. It seems likely to me even further to worsen the unbalance of our present plans. [Oppenheimer is referring to the program to make ever more efficient and powerful fission weapons.] What does worry me is that this thing appears to have caught the imagination, both of the congressional and military people, as the answer to the problem posed by the Russian advance. It would be folly to oppose the exploration of this weapon. We have always known it had to be done, though it appears to be singularly proof against any form of experimental approach. But that we become committed to it as the way to save the country and the peace appears to me full of dangers.

"We will be faced with all this at our meeting; and anything that we do or do not say to the President, will have to take it into consideration. I shall feel far more secure if you have had an opportunity to think about it."

When I interviewed him, Rabi described the meeting of the advisers, of which he was one. He said, "It was a real crisis time, and the whole problem was so entangled that it is difficult for me to remember all the ins and outs of our debate. In fact, the debate was recorded but the tapes were deliberately destroyed soon afterward. It is a great pity that we cannot hear the voices of Fermi and Oppenheimer and the rest in that fateful discussion." Then he explained, "The problem that came up was this: All through the period at Los Alamos there had been a proposal for a bomb called the Super, on which work was being done. The general idea was to cause a fusion reaction involving hydrogen. It was an interesting idea, except that whatever specific theoretical proposal was made didn't seem to produce a self-propagating chain reaction. Beautiful

and clever calculations were made about it, and even more brilliant calculations to show that it wouldn't work. It began to be
more and more evident that if there was such a device, it would
have to be enormous. A ship would be needed to hold it. And then
there was the question of how it would be set off. Would the fusion
process spread—'propagate'—throughout the nuclear fuel or simply confine itself to a small part of it? Would it cool off after you
got it ignited? Since it would be at very, very high temperature,
would the device radiate away that heat and cool off before it could
explode? Nevertheless, after the Russian explosion, certain people
thought we did need some counterthrust. And Lawrence, Alvarez
[Luis Alvarez was a colleague of Lawrence's at Berkeley, who won
the Nobel Prize for Physics in 1968], and Teller felt that the only
thing to do was to go full tilt for this Super. In fact, during the Los
Alamos period some people felt that it should have a higher priority than the ordinary fission bomb. 'We shouldn't dillydally with
the fission bomb but go for the Super'—so said some very eminent
physicists. Not very sensible, but very eminent. I wonder what they
think of that now. Anyway, after the Russian explosion they wanted
to go ahead with a specific Super model. This model is what came
up before the committee. There were two objections to it. In the
first place, it was a very chancy thing, because basically, we didn't
know how to make it. And then, just about this time, it was shown
that it wouldn't work, it wouldn't propagate—at least not that particular model. But the general kind of thing they were talking
about would have been absolutely devastating if it worked, because it was so big. Well, the Super people pressed the Atomic Energy Commission to give this program first priority. They wanted
to try various configurations of this general type. There was no way
we had of refuting these models in general. A man could come in
with some configuration and be strongly in favor of it, and it might

take three months or more of calculations, by some very brilliant people, to find out if it had a chance. There were no experiments, and the constants weren't known, or anything of that sort. It was like dealing with some of the perpetual-motion people. You show them why something doesn't work and they say, 'Fine. Thank you,' and the next day they're back with a modification."

After their deliberations, the members of the Advisory Committee in attendance, all eight, voted unanimously against a crash program for the "super." Seaborg was not there, so his vote was not recorded. The majority report was signed, and certainly written, by Oppenheimer. It is dated October 30. In part it states,

"We base our recommendations on our belief that the extreme dangers to mankind inherent in this proposal wholly outweigh any military advantage that could come from this development. Let it be clearly realized that this is a super weapon; it is in a totally different category from an atomic bomb. The reason for developing such super bombs would be the capacity to devastate a vast area with a single bomb. Its use would involve a decision to slaughter a vast number of civilians. We are alarmed as to the possible global effects of the radioactivity generated by the explosion of a few super bombs of conceivable magnitude.* If super bombs will work at all, there is no inherent limit in the destructive power that may be attained with them. Therefore, a super bomb might become a weapon of genocide."

An important technical point is implicit here. There is a limit—part practical and part physics—to the size of a fission bomb. Fundamentally, the limit is set by the fact that if the component parts exceed the critical mass, they will detonate before as-

---

*This somewhat peculiar locution presumably means with a bomb of unlimited magnitude.

sembly.* This means that a pure fission bomb can never exceed in yield a certain maximum. The largest pure† fission bomb ever tested was the so-called Ivy King, which exploded in an air drop on the Eniwetok Atoll on November 15, 1952. This was a uranium bomb whose composition was as close to the critical mass as possible, consistent with safety. The yield was five hundred kilotons—half a megaton. On the other hand, as I have mentioned, hydrogen bombs operate in the million-ton range, and in principle there is no limit to their size. There is no critical mass. Freeman Dyson has pointed out to me the irony in all of this. Over the course of years the military, both here and in the Soviet Union, came to agree that for any military purpose a five-hundred-kiloton bomb was large enough. But these, as the Ivy King test showed, could be made of uranium and use conventional fission techniques. Thus even if the hydrogen bomb had never been invented, the nuclear stockpile would have been about the same as it is at present since the larger hydrogen bombs have been eliminated.

The Advisory Committee report goes on, "The existence of such a weapon in our armory would have far-reaching effects on world opinion: reasonable people the world over would realize that the existence of a weapon of this type whose power of destruction is essentially unlimited represents a threat to the future of the human race which is intolerable. Thus we believe that the psychological effect of the weapon in our hands would be adverse to our interest.

*There is a subtlety here that I am skipping over. The critical mass depends on the density of the material. When the plutonium is compressed, its density is increased and the critical mass decreased. Thus you can have material that exceeds the critical mass at the higher density which only goes critical when compressed.

†There have been larger fission bombs tested, but these were "boosted."

"We believe that a super bomb should never be produced. Mankind would be far better off not to have a demonstration of the feasibility of such a weapon until the present climate of world opinion changes.

"It is by no means certain that the weapon can be developed at all and by no means certain that the Russians will produce one within a decade. To the argument that the Russians may succeed in developing the weapon, we would reply that our undertaking it will not prove a deterrent to them. Should they use the weapon against us, reprisals by our large stock of atomic bombs would be comparably effective to the use of a super.

"In determining not to proceed to develop the super bomb, we see a unique opportunity of providing by example some limitations on the totality of war and thus of limiting the fear and arousing the hopes of mankind."

The minority report was signed by Fermi and Rabi. It was even more blunt. They write, for example, "Necessarily such a weapon goes far beyond any military objective and enters the range of very great natural catastrophes. By its very nature it cannot be confined to a military objective, but becomes a weapon which in practical effect is almost one of genocide." Rabi told me, "In any event, there was strong agreement within the committee that we should not go ahead. We all agreed that if it could be made to work it would be a terrible thing. It would be awful for humanity altogether. It might give this country a temporary advantage, but then the others would catch up—and it would just louse up life. Fermi and I said that we should use this as an excuse to call a world conference, for the nations to agree for the time being, not to do further research on this.

"Fermi and I felt that if the conference should be a failure and we couldn't get an agreement to stop this research and had to go ahead, we could then do so in good conscience. Some of the others, notably Conant, felt that no matter what happened it shouldn't be made. It would just louse up the world. So the committee unanimously agreed not to go ahead with it with this high priority, especially since we were doing very well with the development of the fission bomb. One member was absent—Glenn Seaborg—and I don't know how he would have voted if he had been there for the discussions. Well, the committee's report to the AEC caused a lot of flak—a lot of consternation—in some circles, and especially with the Berkeley group and Teller. I think we persuaded some members of the commission, but one, Lewis Strauss, got into an absolute dither. The Berkeley people and Strauss went around and talked to newspapermen all over, and to the House and Senate, and whatnot. Furthermore, it was just about then that Klaus Fuchs was arrested in London for espionage on behalf of the Soviet Union."

Rabi continued, "All of this got to President Truman and built up such a head of steam that he was practically forced to declare that he was going to give the Super top priority." On January 31, 1950, Truman announced, "Accordingly I have directed the Atomic Energy Commission to continue its work on all forms of atomic weapons, including the so-called hydrogen or superbomb." The advisers had agreed that there should be a program to study the hydrogen bomb—not a crash program—but the difference was that now there was a declared policy for all to see—above all the Russians—that we were working on the hydrogen bomb. On March 10, in a secret order, Truman directed the AEC to expand its hydrogen bomb program. Reacting to the earlier Truman declaration, Lavrenty Beria, Stalin's notorious secret police commander, whom he had placed in charge of the weapons program, put that program on a crash basis.* The race for the hydrogen bomb was on.

Rabi commented, "Now around this very time, some experiments that were suggested by Teller and Ulam gave rise to a new idea for making thermonuclear reactions. [Rabi has foreshortened the chronology. The Ulam-Teller idea came a year later. In the interim the crash program was focused on the "super" that no one knew how to make.] It bore no relation to the original super, and in a fairly short time it was shown to be a practical thing, which could be calculated. And the General Advisory Committee backed it—with reluctance, because of all the problems it would create. But it was not the horrendous first thing. It was a terrible thing, but not the original Super. It came not long after Truman's original decision, so the two devices got mixed up in people's minds. Now, I never forgave Truman for buckling under pressure. He simply did not understand what it was about. As a matter of fact, after he stopped being President he still didn't believe that the Russians had a bomb in 1949. He said so. So for him to have alerted the world that we were going to make a hydrogen bomb at a time when we didn't even know how to make one was one of the worst things he could have done. It shows the dangers in this kind of thing. He didn't have his own scientific people to consult and give him impartial advice."

Clearly, the Ulam-Teller idea changed the whole debate. In the 1954 hearing, Oppenheimer made that clear. He said, ". . . It is my judgment in these things that when you see something that is technically sweet, you go ahead and do it and you argue about what to do about it only after you have had your technical success.

---

*There was already a program in place that had been inspired by information supplied by Fuchs. I thank Norman Dombey for describing Fuchs's role to me.

That is the way it was with the atomic bomb. I do not think anyone opposed making it; there were some debates after it was made. I cannot very well imagine if we had known in late 1949 what we got to know in early 1951 [the Ulam-Teller proposal] the tone of our report would have been the same. You may ask other people how they feel about that. I am not sure they will concur; some will and some will not." Of course, we cannot rewrite history, so we don't know how the advisers would have voted if they had had this "technically sweet" idea in front of them. We can only wonder if those of them who were so outspoken when they argued against the development of the hydrogen bomb on moral grounds, at a time when it appeared impossible to make, would have been as passionate if it had been clear then that the bomb could be made. As Rabi said, upon learning of the new idea the General Advisory Committee changed its advice and backed the development of the hydrogen bomb.

The Ulam-Teller proposal is thus so important, and there has been so much confusion about it, that it demands explanation. It was highly classified at the time, and I must reiterate what I said earlier about classified material. My only contact with these matters was in the late 1950s and early 1960s. I spent a summer at Los Alamos and then briefly consulted for the Rand Corporation. I consulted more extensively for General Atomics on another project invented by Ulam, a nuclear-powered space ship called the Orion. I had at this time what is called a "Q clearance"—the kind that Oppenheimer lost—which entitled me to know whatever I needed to know for whatever I was working on. Since none of my work involved hydrogen bombs, there was never any need for me to know about the Ulam-Teller proposal. In later years a number of articles and some excellent websites have appeared where these matters are gone into. I have relied heavily on them. First, a few remarks about Ulam.

Stanislaw "Stan" Ulam was born in 1909 in Lvov, Poland. He was one of several brilliant Polish mathematicians who flourished there before the war. He sought refuge in the United States in the late 1930s and was a Junior Fellow at Harvard and a visitor to the Institute for Advanced Study, where he worked with von Neumann. When the war broke out he was teaching at the University of Wisconsin. It was von Neumann who persuaded Ulam to involve himself with the atomic bomb project, though he could not tell Ulam anything about what it was or even exactly where he would be going. Ulam learned that he was to take the train to Lamy, New Mexico. He also noticed that several of his colleagues in physics were disappearing. They were not able to say where, or why. Knowing nothing about New Mexico, Ulam went to the library to get the WPA guide to the state. When he looked at the list of previous borrowers, he noticed that it included his colleagues who had disappeared. For Ulam, the Los Alamos mesa and surrounding countryside were revelations. He had never seen scenery like this. In any event, he was placed in Teller's group. Almost from the beginning the two men disliked each other. Teller makes no secret of this in his memoirs. After the war Teller left Los Alamos while Ulam remained until the late 1960s, when he went to the University of Colorado in Boulder, from which he retired in 1973, returning to Santa Fe where he stayed until his death in 1984. He continued to consult for Los Alamos.

Robert Marshak once said to me that he thought Ulam was a "smart" mathematician, by which he meant that he could do things that were actually useful to physicists. One of the things he did during the war was to invent a kind of statistical sampling method for numerically evaluating otherwise intractable mathematical expressions. He must have thought it resembled a gambling game because it came to be called the "Monte Carlo" method. (A web search reveals that there are some 300,000 sites devoted to it.) In the period we have been discussing, Ulam was busy showing that various of Teller's schemes for making the "super" would not work. He was one of the people that Rabi said were doing "brilliant" calculations. It was only in this sense that he was working on the hydrogen bomb. He was, however, interested in making more powerful fission bombs. In this context he invented, or at least revived, an idea that became known as "super-compression."

It will be recalled that a plutonium fission bomb is made to detonate by compressing, say, a sphere of plutonium with high explosives so that the density of the sphere is increased to the point where the critical mass is exceeded. Ulam wanted to replace, or augment, this use of high explosives, which seemed inefficient. Thus he thought of putting two atomic bombs in proximity. The first bomb would be exploded conventionally, but the idea was to make use of the kinetic energy from the particles—such as the neutrons—produced in the first explosion, to compress the plutonium sphere of the second bomb. You would then have a two-stage weapon. Someone has remarked that the explosive energy sufficient to flatten a city like Hiroshima is being used in this scheme to compress a sphere of plutonium.

For the moment, this had nothing to do with hydrogen bombs. Ulam had this idea in December 1950. But by January 1951 he realized that the same idea could be used to compress a container of fusible material, thus heating it so that fusion reactions would take place. By the end of January, Teller became involved. He made a crucial suggestion. It was based on the fact that at the high temperatures of a fission bomb, at least 80 percent of the energy is given off in the form of soft x-rays, which move at the speed of light.

To explain how the Ulam-Teller idea works, I refer to the figure below, again taken from the web. Because all the original Ulam-Teller documents are classified, I cannot say what stage of their work this diagram represents. Nor can I say who had which idea. Teller's version, expressed again in his memoir, was that all the ideas were his. He must have been very frustrated when an individual he didn't like—Ulam—was able to do something he had failed to do for nearly a decade.

In this figure the circles on the left represent the primary bomb with its high-explosive implosion. The x-rays it produces extremely

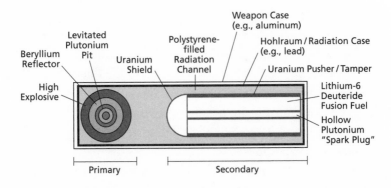

rapidly fill the radiation channel that surrounds the secondary bomb. The polystyrene is converted into a white hot plasma. The way the secondary bomb is imploded is quite subtle. The surface of what is labeled the uranium pusher/tamper is heated by the plasma and expands. This surface is then blown off. This rush of material acts like rocket fuel being expelled from the tail of a rocket. It is the reaction momentum from the stream of ablated material that implodes the secondary bomb. This implosion causes the cylindrical rod of plutonium—the "spark plug"—to go critical. There are so many neutrons around from the first fission bomb that a second fission chain reaction takes place in the plutonium. But this generates enough heat to raise the temperature of the fusible material around the spark plug so that fusion reactions can now take place. The "tamper" keeps the radiation from the high-temperature fusible material from escaping and thus shutting off the fusion reactions. If uranium is used in the tamper—as shown—the neutrons from the fusion reaction will induce a fission chain reaction in the imploding tamper. This produces about half the energy released in the bomb. Thus the sequence is fission-fusion-fission, with most of the energy from a hydrogen bomb actually coming from fission.

As I have said, I do not know what version of this was presented to the Advisory Committee which, as Oppenheimer noted, found it "technically sweet." What strikes me every time I study this is the incredible human ingenuity that went into it, both in the United States and presumably in the other countries that followed in our footsteps—ingenuity employed in the service of taking as many human lives as possible.

Once the hydrogen bomb project started at Los Alamos, Teller decided he could not work with the people who were now successfully running the program. He left to return to the University of Chicago. The bomb was designed and tested by Los Alamos scientists. Like many other fathers, Teller had deserted his offspring when the heavy lifting became apparent.*

---

*I used to refer to Ulam, who was a good friend, as the mother-in-law of the hydrogen bomb.

Meanwhile Oppenheimer had his own problems. On July 7, 1949, he had been called to testify before the House Un-American Activities Committee chaired by the New Jersey congressman C. Parnell Thomas. One of the committee members was Richard Nixon. The committee was investigating what were alleged to have been subversive activities at Berkeley. They wanted to learn from Oppenheimer what he knew, especially since some of the allegations concerned his students and his brother. Oppenheimer's performance in front of the committee was one of his most disgraceful episodes. Later he offered the excuse that he had been assured his testimony would not be made public—as if that mattered. He made statements about four of his students, but the most damaging concerned one named Bernard Peters.

Peters's story was quite remarkable. He had been born in 1910 in what is now Poznan, Poland, but was then Germany. He eventually studied electrical engineering in Munich but was arrested and sent to Dachau for his anti-Nazi agitation. The Communists had agitated against the Nazis, but Peters denied any involvement with the Communist party. Somehow he escaped from Dachau and traveled to Italy by bicycle to join his eventual wife, Hannah Lilien. They emigrated to the United States, where Peters went to work so he could finance his wife's medical education. When they moved to the San Francisco area, she obtained a research job at Stanford. It would appear that she joined the Communist party in California and became a friend of Jean Tatlock's. This seems to be how Oppenheimer first met Peters. Oppenheimer was so impressed that he took him on as his graduate student, even though Peters did not have an undergraduate degree. After Oppenheimer became the director at Los Alamos, he tried unsuccessfully to persuade the couple to join him, since they offered two very valuable professional skills.

What is incomprehensible is that in 1943, when Oppenheimer was interviewed by a security officer about his students and possible security problems at Berkeley, he labeled Peters as a "crazy person" and potentially dangerous. He seems to have based this on the fact that Peters took part in anti-Nazi riots which, Oppenheimer claimed, falsely, he did as a Communist. In the 1949 hearings he was again asked about this and now repeated his claims with adornments. Oppenheimer's testimony, quoted verbatim, was leaked to the *Rochester Times Union*, chosen because Peters then had a job at the University of Rochester. Several things happened. Needless to say, Peters was outraged. Eventually he confronted Oppenheimer, who agreed to try to repair the damage. Oppenheimer also received strong letters of protest from old friends like Hans Bethe and Victor Weisskopf. He wrote a letter, which was published in the *Times Union*, in an attempt to retract what he had said. To the great credit of the University of Rochester, and its then president, the historian Alan Valentine, it not only refused to fire Peters but in fact promoted him to full professor. The damage was done, however, by the government. Peters was a cosmic ray experimental physicist and needed to do research in various high-altitude stations around the world. But Peters's American passport was taken away. He left the United States and moved first to India and finally to the Niels Bohr Institute in Copenhagen, where he remained until his death in 1993. During his 1949 HUAC hearing, Oppenheimer so ingratiated himself with the committee that when they began to query him about his brother, he got them to back off. At the end, Congressman Nixon congratulated him and wished him luck.

Two more names now resurfaced: Ernest Lawrence and Oppenheimer's greatest vulnerability from the past—Haakon Chevalier. Both were fraught with unresolved issues. First Chevalier.

Perhaps the best way to start is with Chevalier's version of what happened. This is contained in his book *Oppenheimer: The Story of a Friendship*. He writes,

"George and Dolly Eltenton were an English couple who lived in Berkeley with their three children. We [Chevalier and his wife] had first met Dolly about 1937, when she had come to the office of the League of American Writers [a Communist-dominated organization to which many distinguished writers, Communist and non-Communist, belonged] in San Francisco to volunteer to do secretarial work for us, and later she introduced her husband to us. She was, we learned, a first cousin of Sir Hartley Shawcross [a British jurist and Labour parliamentarian who died in July 2003 at the age of 101; he had been chief prosecutor at the Nuremberg trials and later at the trial of Klaus Fuchs], and her husband was a chemical engineer who worked for the Shell Oil Company. They were, both of them, intelligent, cultivated and sociable. They had lived for several years in Leningrad where George had had a job with a British firm, and there one or two of their children had been born. We had seen Dolly often at League meetings during the period when the League was more or less active, and we met the two of them occasionally at the houses of common friends or at fund-raising parties."

Having set the stage, Chevalier now comes to the "incident." He notes that sometime late in 1942 or early 1943—no one involved in any of this is sure of the dates—Eltenton asked to see him at his house. Chevalier writes,

"Why had he asked to see me? He began with some general remarks on the war—the uncertainty, the odds against which the Allies were fighting, and the fact that the Soviet Union and the United States were brothers-at-arms and were bearing the responsibility for winning the war. He then spoke of the importance of the work being done by American and Soviet scientists. Their role could be decisive. [It does not seem to have occurred to Chevalier to wonder how Eltenton knew this. Indeed, what work was he talking about?] The Soviet scientists, he said, felt that in order to make the most telling use of the scientific know-how and resources of both countries for the combined war effort it was highly desirable that there be close collaboration between the scientists of both countries, as there was in other fields, so that each could benefit by the work being done by the other. Oppenheimer was known to be in charge of an important war project, and was also known to be very much of a left-winger. [This is presumably a paraphrase of what Eltenton is telling Chevalier, which raises the same questions as before. Oppenheimer had recently been made the director of Los Alamos. Did Eltenton know this, and if so, how? Chevalier did not know it.] He was therefore likely to be sympathetic to the idea of closer scientific coordination and because of his eminence could be effective in promoting it. Since I was a friend of his, the idea [whose idea?] was that I was to be asked to sound him out as to how he felt about the possibility of such collaboration."

Before I continue with the quotation, it seems to me that Chevalier's story so far refutes one extreme view of Oppenheimer that I will explore more thoroughly when I discuss what actually precipitated the hearing. This was the accusation that Oppenheimer had been a Soviet agent since 1939. If this had been true there would have been no need for this sort of song and dance from Eltenton. Chevalier goes on,

"Eltenton's manner was somewhat embarrassed. He seemed not too sure of himself. Through his roundabout phrases it gradually became clear to me that what the people behind him [Chevalier does not attempt to identify them] were really interested in was the secret project Oppenheimer was working on."

Chevalier then says that when he realized the implications of what he was being asked to do, he gave an "unqualified 'no.'" He tells us that he was not sure if Eltenton understood the implications either. He tells us that Eltenton seemed relieved when the matter was dropped. But, as Chevalier writes,

"The next question was whether or not to tell Opje about this conversation. I had felt fairly convinced, by the time I left him, that Eltenton was not deeply involved in any kind of conspiracy. [He was wrong.] But what about those for whom he was acting? Whoever was behind this was not likely to give up after a first unsuccessful try. ["Whoever was behind this?" Did Chevalier have no suspicions?] There might be other approaches, and these might cause trouble. Should Opje be forewarned? I decided, after much hesitation, and having discussed it with my wife, who agreed, that he should be told. My chief misgiving was over mentioning Eltenton's name. I didn't want to get him into trouble. I could of course merely report the conversation without mentioning his name. But that would not be very helpful, and if the people who approached him were involved in a serious attempt to get information [What would have constituted a nonserious attempt to get information? We are talking about espionage], it was something Opje would know how to handle [whatever that means]. And besides, Opje was the one being from whom I had no secrets, and it would have been inconceivable for me to withhold any part of the story that might have serious consequences for him. A final consideration was that Eltenton had, after all, undertaken to make the approach, and I would of course convey to Opje my feeling that Eltenton had let himself in for this rather naively and would now drop the matter."

We shall shortly discuss Eltenton's "naiveté," but now we come to the "incident." Chevalier writes,

"Not too long after this the Oppenheimers invited us for dinner on Eagle Hill [their home near Berkeley]. When, after the formalities and amenities of welcome, Opje went out to the kitchen to fetch the ice and the mixings for a martini, I followed him and there told him of my conversation with Eltenton.

"This was the molehill out of which the mountain of the so-called 'Chevalier incident' was subsequently built. What was actually said during that conversation? Neither Opje nor I have the slightest recollection. I remember only the substance: that I reported the conversation I had had with Eltenton because I thought he should know of it, and that he agreed I was right in telling him. He was visibly disturbed, we exchanged a remark or two. We went back into the living room with the cocktail shaker, the gin and the vermouth, and joined our wives. I dismissed the whole thing from my mind."

The least one can say is that this account is disingenuous. For a start, both here and throughout his book Chevalier leaves out the fact that he was a member of the Communist party at this time. So were Eltenton and his wife. He also leaves out the fact that a well-established spy network was operating from the Soviet consulate in San Francisco. Possibly Chevalier did not know this in 1943, but he surely must have known it in 1965 when he published his book. By that time it was well known. As for Eltenton, he had had a large number of contacts, social and otherwise, with the two operatives in the consulate—Gregori Kheifets and Pyotr Ivanov—who ran the espionage program. Indeed, Eltenton was an early and willing recruit. In late 1942, Ivanov asked Eltenton to find a way to contact Oppenheimer to learn about the atomic energy work that was going on at the Berkeley Radiation Laboratory, about which the Soviets knew something. Eltenton suggested Chevalier. When Eltenton asked how such information might be transmitted to the

Soviet Union, Ivanov made sounds like the clicking of a camera. This is what Eltenton later told the FBI. We are left to wonder what, if any, of this Chevalier knew in 1943.

The reader will no doubt have, by this time, formulated the question I have been sidestepping: Was Oppenheimer himself a member of the Communist party? Here again we run into a wall. The essence of the situation is this: A certain number of people, including Chevalier, have claimed they attended meetings of a closed unit of the Communist party at which Oppenheimer was present. A year before Chevalier wrote his book, he sent a remarkable letter to Oppenheimer. He informed Oppenheimer that he was well along in the writing of what became his account of the "Chevalier incident." He mentions, almost casually, that he plans to include the "fact" that between 1938 and 1942 he and Oppenheimer were members of the same closed Communist party unit. He tells Oppenheimer that he wants to put this in its "proper perspective" and asks if Oppenheimer will have any objections.

In reading this, one is led to wonder if Chevalier is naive, malicious, or both. He knew, without any doubt whatsoever, that on more than one occasion Oppenheimer had made sworn statements to the effect that he had not been a member of the Communist party. He was now accusing Oppenheimer, in essence, of committing perjury. Would Oppenheimer have any objections? After consulting his lawyers, Oppenheimer responded on August 7, 1964, in a letter in which he once again clearly stated that he had never been a member of the Communist party, something he says he was sure Chevalier must have known. Oppenheimer always claimed that he attended these meetings out of curiosity, not as a member of the party. Chevalier backed off, and his book does not mention the matter. He also does not describe his own connection with the party. As far as I am concerned, this is it. One can believe

Oppenheimer, or one can believe the party members who claim he was one of them. I believe Oppenheimer, and until someone comes up with real evidence, I will continue to do so.

The "Chevalier incident" involved not only this encounter in Oppenheimer's home but what Oppenheimer did about it. Here we are confronted with one of the strangest bits of behavior in the whole Oppenheimer saga. In the first instance, Oppenheimer did nothing about it until August 25, when he returned to Berkeley from Los Alamos for a visit. At this time he paid a brief call on Lieutenant Lyall Johnson, the military officer who was responsible for security for the Berkeley nuclear project. In the course of it he told Johnson that Eltenton was someone worth watching. This was not news to Johnson since they were already watching Eltenton. Obviously Johnson wanted to know how Oppenheimer had come upon Eltenton, but Oppenheimer offered no further information. Johnson reported his conversation with Oppenheimer to his senior officer, Lieutenant Colonel Boris Pash. It was Pash whom Groves would appoint, in the fall of 1944, to lead the reconnaissance mission on the European continent to discover if the Germans had made progress toward building a bomb. But on August 26, Pash convened a second interview with Oppenheimer in the presence of Johnson and a tape recorder. It was the revelations of this interview that many of the principals involved in the 1954 hearings thought did Oppenheimer the most damage. It was presented in its entirety at the hearings, and Oppenheimer was cross-examined on it.

The interview begins amiably enough. Pash tells Oppenheimer that he does not mean to take up much of his time, to which Oppenheimer replies, "That's perfectly all right. Whatever time you choose." Then they get down to business. Pash wants to know about any groups that might be interested in nuclear espionage.

Oppenheimer's reply is the first installment of a story that becomes more and more baroque as the interview proceeds. He says,

". . . I have no first hand knowledge that would be for that reason useful, but I think it is true that a man, whose name I never heard, who was attached to the Soviet consul, has indicated indirectly through intermediary people concerned in this project that he was in a position to transmit, without any danger of a leak, or scandal, or anything of that kind, information which they might supply. I would take it that it is to be assumed that a man attached to the Soviet consulate might be doing it but since I know it to be a fact, I have been particularly concerned about any indiscretions which took place in circles close enough to be in contact with it. To put it quite frankly—I would feel friendly to the idea of the Commander in Chief informing the Russians that we are working on this problem. At least I can see that there might be some arguments for doing that, but I certainly do not feel friendly to the idea of having it moved out the back door. I think that it might not hurt to be on the lookout for it."

Even knowing what we now do, it is difficult to deconstruct this fabulation. There is no mention of Eltenton, to say nothing of Chevalier. Instead the Soviet consulate is brought into the act. One wonders if this is something Oppenheimer invented or whether part of the conversation with Chevalier—the part not reported—might have had to do with the consulate. It is interesting that Oppenheimer sees possible merit at this stage in collaboration with the Russians. Of course Pash was not satisfied with this version and tries to home in. Oppenheimer then offers a second version.

"Well, I might say that the approaches were always to other people, who were troubled by them, and sometimes came and discussed them with me; and that the approaches were always quite indirect so I feel to give more, perhaps, than one name, would be to implicate people whose attitude was one of bewilderment rather than one of cooperation. I know of no case, and I am fairly sure that in all cases where I have heard of these contacts would not have yielded a single thing. That's as far as I can go on that. Now there is a man, whose name was mentioned to me a couple of times—I don't know of my own knowledge that he was involved as an intermediary. It seems, however, not impossible and if you wanted to watch him it might be the appropriate thing to do. He spent quite a number of years in the Soviet Union. He's an English *** I think he is a chemical engineer. He was—he may not be here now—at the time I was with him, employed by the Shell development. His name is Eltenton. I think there was a small chance—well let me put it this way: He has probably been asked to do what he can to provide information. Whether he is successful or not, I do not know, but he talked to a friend of his who is also an acquaintance of one of the men on the project, and that was one of the channels by which this thing went. Now I think that to go beyond that would be to put a lot of names down, of people who are not only innocent but whose attitude was 100-percent cooperative."

In this version we have Eltenton talking to a friend who knew one of the men on the project. Pash persists, and Oppenheimer says of Eltenton's "friend," "It's a member of the faculty, but not on the project." This is as far as Oppenheimer is willing to go, except to add that two other people were contacted within a week of each other—another apparent fabulation. The obvious question is, Why did Oppenheimer invent this fiction? He explained later that he was trying to shield Chevalier, who was his friend, while attempting to alert the authorities to what seemed to be attempted espionage. It is hard to imagine a worse way of going about this—and it doesn't get better.

Pash, who had now been sent on a wild goose chase, reported the matter to General Groves. In early September, Oppenheimer made a sixteen-hour train trip from Cheyenne to Chicago with Groves and his security officer, Colonel John Landsdale. The subject of the Pash interview came up, and Oppenheimer said he would provide the name of the Berkeley professor if Groves ordered it. Incomprehensibly, Groves let the matter drop, but not Landsdale. On September 12, the occasion of Oppenheimer's next trip to Washington, Landsdale conducted a formal interview. This time he tried a different tactic. He proposed several names and asked Oppenheimer to comment on them. When he came to Chevalier, the following dialogue occurred.

L: How about Chevalier?
O: Is he a member of the party?
L: I don't know.
O: He is a member of the faculty, and I know him well. I wouldn't be surprised if he were a member. He is quite a Red.

An exchange like this causes people like myself, who have tried to understand Oppenheimer, to throw up their hands in despair. Oppenheimer must have known that Chevalier was a member of the party. If he wanted to protect him, why did he make this kind of statement? Was he so naive as to think this would not hurt Chevalier? Why not at this point simply give the whole story? And it gets worse.

Groves let things slide until December 12. On a visit to Los Alamos, he ordered Oppenheimer to divulge the name of his contact. What was then actually said is a matter of some dispute. According to Landsdale, Groves told him that Oppenheimer said Chevalier had one contact, with his own brother, Frank. Another version is that Oppenheimer stuck with his story about three contacts. And there is Oppenheimer's version, in which he claimed that he told General Groves he had been Chevalier's sole contact. When questioned at the hearing about his first interview with Pash, the following colloquy occurred. "Q" is the questioner, Roger Robb, the attorney who conducted the hearings on behalf of the Security Board, and "A" is Oppenheimer.

Q. Did you tell Pash that X had approached three persons on the project? [Recall that Oppenheimer had told Pash that some unnamed Berkeley faculty member was the intermediary. So in this interview Chevalier's name had not yet surfaced.]

A. I am not clear whether I said there were three X's or that X approached three people.

Q. Didn't you say that X had approached three people.

A. Probably.

Q. Why did you do that Doctor?

A. Because I was an idiot.

During the hearings Oppenheimer had many defenders, but they were all appalled by the Chevalier affair. The best they could say was that Oppenheimer had been caught in a question of divided loyalties and had made the worst possible choices for himself and everyone else. People who defended him said this kind of lapse had never recurred and that Oppenheimer had learned a lesson from it. His adversaries argued that it showed he was unfit to be a guardian of the nation's secrets. And Chevalier? Nothing happened until June 1946, when he was visited by agents of the

FBI. This in itself is very strange. If the matter was so urgent in 1943, why did the FBI wait until a year after the war had ended to interview Chevalier? In any event, they wanted to know about Eltenton, Oppenheimer, and the three scientists—the three imaginary scientists that Oppenheimer appears to have invented. Needless to say, Chevalier had no useful information to supply, or at least information the agents would accept. From that time on he would be shadowed by FBI agents. During this period he had dinner with the Oppenheimers and told Oppenheimer that under FBI questioning he had revealed the contact with Eltenton and their kitchen conversation. Oppenheimer seemed much upset, but he did not tell Chevalier that he had been the source of this information. This Chevalier did not learn until the details of the hearings were released in 1954. By this time Chevalier, whose life in the United States had been turned upside down by these accusations, had moved to Paris. I think his true feelings about Oppenheimer then were reflected in the letter he wrote to him on July 23, 1964, when he told Oppenheimer that he, Oppenheimer, had nothing to be ashamed of by having been a member of the Communist party. He was sure, he said, that Oppenheimer would have no objections to having this revealed in a book. Now to Lawrence.

I have noted that Lawrence and Oppenheimer came to Berkeley as young professors at about the same time, lived in the Faculty Club, and, until Lawrence's marriage, double-dated. Afterward Oppenheimer became one of the family. In the mid-thirties Oppenheimer became politically active, especially in left-wing causes—something that Lawrence referred to as "leftwandering." One of these causes was the unionizing of Lawrence's Radiation Laboratory in 1941, and one of the representatives of the union involved was Eltenton. Oppenheimer had not bothered to inform Lawrence of his advocacy, and when Lawrence learned of it there was a considerable row. The following year Oppenheimer was made director of Los Alamos, even though Lawrence had strongly lobbied for his protégé Edwin McMillan. At the end of the war he and Oppenheimer had another row about Oppenheimer's returning to Berkeley. Lawrence seemed to feel betrayed when Oppenheimer didn't. And then came the matter of the hydrogen bomb. Lawrence was firmly in the "super" crash program camp. Typically, Lawrence saw in it the opportunity to build something—this time a reactor that could be used to produce super-heavy hydrogen-tritium. He would do this in collaboration with Luis Alvarez, and the two of them began campaigning for it in Washington. They soon discovered that Oppenheimer would not support them, only fueling Lawrence's anger. To all of this was added Lawrence's feeling that Oppenheimer was "immoral."

Apparently Lawrence was told at a cocktail party that Oppenheimer had had an affair with Ruth Tolman, the wife of the Cal Tech physicist Richard Tolman. He passed this bit of gossip on to Lewis Strauss who in a memorandum to himself noted that Oppenheimer first "earned [Lawrence's] disapproval a number of years ago when he seduced the wife of Prof. Tolman at Caltech. According to Lawrence, it was a notorious affair which lasted long enough to become apparent to Dr. Tolman who died of a broken heart." When I read this, I wondered if it was true. While I make no claim to having a list of Oppenheimer's affairs, it seemed to me that something "notorious" should have, by definition, left a trail. The principals are all dead: Richard Tolman in 1948, his wife Ruth in 1957, and Oppenheimer in 1967. But there are still people who were at Cal Tech—where, incidentally, Lawrence wasn't—in the late 1930s when presumably this was happening. Among them is Robert Christy, who became Oppenheimer's student in 1937. The only affair Christy had heard of is the one that Oppenheimer had with his future wife, Kitty, when she was still married to Dr. Stewart Harrison. This *was* notorious, but Harrison remarried and appears to have led an entirely happy life. Tom Tombrello, the chairman of the physics department at Cal Tech, arrived there in 1961, when the affair between Oppenheimer and Kitty was still being discussed. He talked about the Tolman matter with Marge Lauritsen Leighton, the widow of Tom Lauritsen, a Cal Tech nuclear physicist. When she was first married she lived in the Tolmans' house with them, and was a close friend of both of them and the Oppenheimers. She had never heard of any such affair and doubted that it had ever occurred. Moreover, when Oppenheimer came to Cal Tech to teach prior to the war, both before and after he was married, he stayed with the Tolmans. Tolman did not die of a broken heart. He died of a stroke, and Stewart Harrison was the attending physician.

This might seem to put an end to it, but with Oppenheimer there is no end to anything. While Ruth Tolman destroyed many of her personal papers, a few letters to Oppenheimer do survive. In one of them she writes, "The precious times with you that week and the week before keep going through my mind, over and over, making me thankful but wistful, wishing for more. I was grateful for them, Dear, and as you know, hungry for them too." Later in the letter there is an affectionate reference to her husband. In another letter she writes about a possible arrangement by which they might be able to see each other. She describes driving to San Diego and seeing "the long stretch of beach where the sandpipers and gulls played. Soon I shall see you. You and I both know how it will be." Of one thing one can be sure: if there was an affair, it was very private and hardly one that would be common knowledge at cocktail parties.

In any event, Lawrence would have been delighted to put a

long nail into the coffin of Oppenheimer's reputation at the 1954 hearing. But he was too good a scientist-politician not to worry that he might, at the same time, nail shut his own coffin. He understood—which Teller, for example, didn't—that if he testified against Oppenheimer he ran the risk of being ostracized by a large part of the scientific community. His concern, as usual, had an element of self-interest. He realized that if he made enemies of the wrong people he might lose the funding for some of his pet projects. Nonetheless his animus against Oppenheimer was such that he decided to testify anyway. On the way to Washington he stopped off at Oak Ridge for a meeting, and there he had an attack of colitis that was serious enough to cause him to return to California. Before he returned, however, he had a violent telephone confrontation with Strauss, who called him a coward for not coming to Washington to testify. If one were writing a novel about these people, one might speculate that some residue of what Lawrence once felt for Oppenheimer was at work in preventing him from delivering the mortal blow. In any event, Lawrence died from colitis in 1958.

We have now examined some ingredients of the witches' brew out of which the 1954 Oppenheimer security hearings were concocted, namely, his capacity for making enemies of people who could do him harm, and his vulnerability because of his radical past. Two more ingredients need scrutiny: the air force, and the ambiance. The latter was reflected in the rising influence of Senator Joseph R. McCarthy, and the former by Oppenheimer's alleged foot-dragging on the hydrogen bomb. First, the air force.

Major General Roscoe Charles ("Bim") Wilson began his testimony at the hearing with a brief statement: "Mr. Chairman, I would like the record to show that I am appearing here by military orders, and not on my own volition." The orders came from the chief of staff of the air force. Wilson was to represent the views of the air force at the hearings. He was a good choice. He had been flying for the air force since 1929 and had been selected in 1943 to be the air force's liaison officer to General Groves. Wilson's responsibility was to make sure that planes and their crews were ready to drop atomic bombs on Japan. In other words, he was familiar with the nuclear weapons program from the beginning. He was plainspoken and put forth the air force's case in no uncertain terms. He saw in Oppenheimer's reluctance to approve of various things the air force wanted—larger bombs, a nuclear-powered airplane, and the like—"a pattern of action that was simply not helpful to national defense." When asked to explain, he replied,

"I would like first to say that I am not talking about loyalty. I want that clearly understood. If I may, I would like to say that this is a matter of my judgment versus Dr. Oppenheimer's judgment. This is a little embarrassing for me, too. But Dr. Oppenheimer was dealing in technical fields and I was dealing in other fields, and I am talking about an overall result of these actions.

"First I would like to say, sir, that I am a dedicated airman. I believe in a concept which I am going to have to tell you or my testimony doesn't make sense.

"The U.S.S.R. in the airman's view is a land power. It is practically independent of the rest of the world. I feel it could exist for a long time without sea communications. Therefore, it is really not vulnerable to attack by sea. Furthermore, it has a tremendous source of manpower. If you can imagine such a force, it could probably put 300 to 500 divisions in the field, certainly far more than this country could put into the field. It is bordered by satellite countries upon whom would be expended the first fury of any land assault that would be launched against Russia, and it has its historical distance and climate. So my feeling is that it is relatively invulnerable to land attack."

Wilson continues, "Russia is the base of international communism. My feeling is that the masters in the Kremlin cannot risk the loss of their base. The base is vulnerable only to attack by air power. I don't propose for a moment to say that only air power should be employed in case of a war with Russia, but I say what strategy is established should be centered around air power.

"I further believe that whereas air power might be effective with ordinary weapons, that the chances of success against Russia with atomic weapons or nuclear weapons are far, far greater."

Then he comes to Oppenheimer: "It is against this thinking that I have to judge Dr. Oppenheimer's judgments. Once again his judgments were based upon technical matters. It is the pattern I am talking about.

"I have jotted down from my own memory some of these things that worried me.

"First was my awareness of the fact that Dr. Oppenheimer was interested in what I call the internationalizing of atomic energy, this at a time when the United States had a monopoly, and in which many people, including myself, believed that the A-bomb in the hands of the United States with an Air Force capable of using it was probably the greatest deterrent to further Russian aggression. This was a concern."

At this point the transcript indicates with asterisks that classified material is being discussed. I will come back to the matter of classified material when I discuss the hearing. Wilson continues his case against Oppenheimer:

"Dr. Oppenheimer also opposed the nuclear-powered aircraft. His opposition was based on technical judgment. I don't challenge his technical judgment, but at the same time he felt less strongly opposed to nuclear powered ships. The Air Force feeling was that at least the same energy should be devoted to both projects."

The idea of a nuclear-powered airplane, with the shielding this would require, is truly bizarre.

More asterisks, and Wilson concludes: "The approach to the thermonuclear weapons also caused some concern. Dr. Oppenheimer, as far as I know, had technical objections, or let me say, approached this conservatism for technical reasons, more conservatism than the Air Force would have liked.

"The sum total of this, to my mind, was adding up that we were not exploiting the full military potential in this field. Once again it was a matter of judgment. I would like to say that the fact that I admire Dr. Oppenheimer so much, the fact that he is such a brilliant man, the fact that he has such a command of the English language, has such national prestige, and such power of persuasion, only made me nervous, because I felt if this was so it would not be to the interest of the United States, in my judgment. It was for this reason that I went to the Director of Intelligence to say that I felt unhappy."

This feeling of "unhappiness" was the common air force theme. Some felt that Oppenheimer was "loyal" but misguided; few felt that he was actually a Russian spy. All agreed that he must go. He must no longer be allowed to have any influence on military policy. For this, Lewis Strauss, now chairman of the Atomic Energy Commission, was a sympathetic audience. Now, finally, to ambiance.

If one did not live through the McCarthy period of the 1950s, it is difficult to convey what it was like. As it happened, on January 15, 1954, a few months before the Oppenheimer hearings, I attended a hearing of the Permanent Subcommittee on Investigations of the Committee on Government Operations of the United States Senate, Senator Joseph McCarthy's committee, in Boston. I was then a graduate student in physics at Harvard, and one of my teachers, Wendell Furry, was the star witness.

I had taken introductory physics with Furry as a sophomore. It was only much later that I learned he had been one of Oppenheimer's postdoctoral fellows. Furry looked to me like a sort of country bumpkin. He had a large stomach that hung over pants that I thought he held up by suspenders and a rope belt, and he spoke with a sort of drawl. I may misremember the details, but that was the impression. If someone had asked me to have named the person least likely to have been a Communist in the Harvard physics department, I would have put Furry at the top of the list, which goes to show how little one knows about other people. But he had joined the party in 1938 and had remained for several years. I am not sure when he quit, but in 1954 he was no longer a member.

At the time of the McCarthy hearing I was taking a reading course with Furry on electrodynamics, and I guess I must have gone to the hearing to offer moral support. The university had taken a position that put Furry in grave legal jeopardy. They insisted that he be forthcoming about his own activities but that, in the jargon of the time, he would not have to "name names"—tell on others. We have seen that Oppenheimer quite readily named names in his hearing before the House Un-American Activities Committee. The legal jeopardy for Furry was that by refusing to name names, he lost his Fifth Amendment immunity and could be cited for contempt of Congress—which he was, a case that was eventually dismissed. Furry's refusal to name names put McCarthy into a sort of frenzy, or at least he appeared to be in a frenzy. How much McCarthy really believed in what he was saying is an open question. Here is what he said to Furry:

"This, in the opinion of the chair, is one of the most aggravated cases of contempt that we have had before us, as I see it. Here you have a man teaching at one of our large universities. He knows there were six Communists handling secret government work, radar work, atomic work. [This is one of the inventions McCarthy specialized in: fabricated lists.] He refuses to give either the committee or the FBI or anyone else the information which he has. To me it is inconceivable that a university would keep this type of creature on teaching our children. Because of men like this who have refused to give the government the information which they have in their own minds about Communists who are working on our secret work, many young men have died in the past, and if we lose a war in the future it will be the result of the lack of loyalty, complete [immorality][unmorality] of these individuals who continue to protect the conspirators."*

As Furry was leaving the hearing room, a woman spat on him.

McCarthy no doubt would have gone after Oppenheimer, but the AEC hearings preempted him. As it happened, the Army-McCarthy hearings, which destroyed McCarthy, were going on at just about the same time as the Oppenheimer hearings. Nonetheless, on September 15, 1953, McCarthy had held a closed hearing at which the star witness was one Paul Crouch. Crouch was one of those people who seem to crawl out of the woodwork at times like these. While in the army in Hawaii, he was court-martialed for trying to foment a revolution and sentenced to forty years in the Alca-

---

*The hearing transcript must have had some problems with "immorality" and "unmorality."

traz penitentiary. This was reduced to three, and Crouch joined the American Communist party and went to Russia. There, allegedly, he conferred with officers in the Red Army about his plans to infiltrate the American army. After the war he was in a good position to receive money from various of our government agencies for his information, such as it was.

In his hearing before the McCarthy Committee, Crouch claimed that Oppenheimer was a member of the party and that he had attended a closed party meeting in Oppenheimer's home. This later drew the attention of the AEC and became an element in the hearing. Oppenheimer had to demonstrate that he was not in Berkeley at the time of the supposed meeting. On top of this, the Republicans had made it part of their platform for the 1952 presidential election that they would root out Communists in government. Putting all of this together, there was an explosive mixture simply waiting for someone to light the match. The man who did it turned out to be a Yale graduate, a lawyer, and a lifelong Democrat who actually disliked McCarthy, named William Liscum Borden.

Borden had served in Europe during the war as a bomber pilot. In returning from one of his missions he was alarmed when an object streaked by his airplane. It turned out to be a German missile on its way to England. Borden immediately realized that something new had entered warfare, especially if you combined rockets with nuclear weapons. In 1946 he published a highly influential book, *There Will Not Be Time*, about the prospects of missile warfare. It caught the attention of Senator McMahon, chairman of the Joint Committee on Atomic Energy. He made Borden the committee's executive director. In this capacity Borden was witness to the hydrogen bomb debate and was strongly opposed to people like Oppenheimer who he felt were dragging their feet. Moreover,

Borden had access to the AEC security files. When he read Oppenheimer's, he decided there was an obvious explanation for his behavior: Oppenheimer was a Soviet agent.

In November 1953, Borden produced a letter, which he sent, registered, to J. Edgar Hoover, outlining in detail his charges against Oppenheimer. Since it was based on the AEC security files, it contained mainly information that had been familiar to Hoover when he cleared Oppenheimer in 1947. The new material had to do with things like Oppenheimer's opposition to a hydrogen bomb crash program and his lack of enthusiasm for making nuclear-powered aircraft. The conclusion was that Oppenheimer had been, and still was, an espionage agent.

Borden's letter arrived at the Justice Department and was read by Hoover, who, on November 30, sent a report, along with the letter, to the White House with copies to Strauss, who was now chairman of the AEC, and Charles Wilson, who was then secretary of defense. On December 3, Strauss was summoned to the White House to attend a meeting that had been called by President Eisenhower to discuss the Borden letter. The president wanted to know if the AEC had held a formal hearing on Borden's charges. Learning that it had not, Eisenhower ordered a "blank wall" to be placed between Oppenheimer and any classified material. Agents removed all of this material from Oppenheimer's office at the Institute for Advanced Study, a process that had already begun the previous July.

While this was going on, the Oppenheimers were in Europe. He gave the Reith Lectures on the BBC and visited Paris, where he and his wife had dinner with the Chevaliers. (This also became an issue in the hearings.) When he returned to Princeton he found an urgent message to call Strauss. A meeting was set up in Washington on December 21, 1953 which was also attended by Nichols, the AEC general manager. It will be recalled that it was Nichols who wrote the December 23 letter that charged Oppenheimer with being a security risk. The draft letter was eight pages long and was handed to Oppenheimer to read. It contained the full budget of charges and ended with the statement that the AEC was suspending Oppenheimer's clearance.

As a practical matter, Oppenheimer had been doing very little consulting for the AEC in the years before the hearing. His contract had been scheduled to expire on June 30, 1953, but had been extended for a year, meaning that it would expire on June 30, 1954. That the contract was still in force is what gave the AEC the jurisdiction to conduct its hearing. But the agency had to conduct it by June 30, because afterward Oppenheimer would no longer be an employee. The AEC could simply have let his contract expire without a hearing. That this was not done was addressed in the majority report of the Personnel Security Board, the three-man entity that was set up to conduct the hearings. This was the report explaining the majority opinion to deny Oppenheimer his security clearance. As we shall see, the panel voted two to one to suspend clearance. The authors of the majority report wrote,

"In good sense, it could be recommended that Dr. Oppenheimer simply not be used as a consultant, and that therefore there exists no need for a categorical answer to the difficult question posed by the regulations [regarding clearance], since there would be no need for access to classified material.

"The Board would prefer to report a finding of this nature. We have had a desire to reconcile the hard requirements of security with the compelling urge to avoid harm to a talented citizen."

Then comes the argument:

"The Board questioned why the Commission [the AEC] chose to revoke Dr. Oppenheimer's clearance and did not follow the alternative course of declining to make use of his services assuming it had serious questions in the area of security. To many this would seem the preferable line of action. [They may be referring to Rabi's testimony, quoted earlier.] We think the answer of the Commission to this question is pertinent to this recital. It seemed clear that other agencies of the Government were extending clearance to Dr. Oppenheimer on the strength of AEC clearance, which in many quarters is supposed to be an approval of the highest order. Furthermore, it was explained that without the positive act of withdrawal of access, he would continue to receive classified reports on Atomic Energy activities as a consultant, even though his services were not specifically and currently engaged. Finally, it is said that were his clearance continued his services would be available to, and probably would be used by, AEC contractors. It is noted that most AEC work is carried on by contractors. Withdrawal of clearance and Dr. Oppenheimer's request for a hearing precipitated this proceeding."

In other words, the specter of Oppenheimer's consulting for someone, anyone, on atomic energy was too frightening to be allowed to continue.

In the December 21 meeting with Strauss and Nichols, Oppenheimer was told he had two choices. He could elect to have a hearing before a three-man panel selected by the AEC to contest the suspension of his clearance, or he could resign, which would stop the process. Oppenheimer asked how long he had to make up his mind, and was told he would have to respond by the next day. Upon leaving Strauss's office, Oppenheimer went to the law offices of Joseph Volpe, who had been the AEC's general counsel. They were joined by Herbert Marks, an attorney and one of Oppenheimer's closest friends. Unknown to them, Volpe's office had been bugged, so that Strauss and his allies were privy to this private conversation. This sort of illicit activity continued throughout the hearing, and indeed, despite FBI cautions against doing this, Strauss passed along information from this eavesdropping to the AEC's counsel. The next day Oppenheimer was still wrestling with how to respond. Finally, Oppenheimer wrote a letter to Strauss in which he said, "I have thought most earnestly of the alternative suggested. Under the circumstances this course of action would mean that I accept and concur in the view that I am not fit to serve this government that I have now served for some 12 years. This I cannot do." The stage was set. Both sides began preparing for the now inevitable hearing.

Oppenheimer's immediate task was to find counsel to represent him. His first choice was Volpe, who declined since he thought having been general counsel to the AEC might represent a conflict of interest. If Volpe had agreed, I think the hearings would have been quite different. I do not know if the outcome would have been different, but the hearings would have been much more confrontational. I base this on the advice that Volpe gave while the hearings were in progress. When he heard the conditions under which Oppenheimer and his lawyers had been forced to operate—I do not refer here to the bugging, which Volpe did not know about, but to other matters that I will explain—his advice was for them to throw down the gauntlet, to insist that the playing field be leveled otherwise they would walk out and presumably make a public declaration of what was happening. This might have worked. The panel members constantly tried to explain how fair they were as they made one prejudicial ruling after another. They also expressed concern with how the hearings might be perceived by the scientific community at large. They kept pointing out that what they were doing was no different in principle from what Lilienthal had done in 1947, when he and the other AEC commissioners vetted Oppenheimer's clearance. Perhaps they were being sincere, but a tough lawyer like Volpe would have cut through all this like a knife going through melted butter.

In the event, in January Oppenheimer selected Lloyd K. Garrison to be his counsel. He was the great-grandson of the abolitionist William Lloyd Garrison. Like his great-grandfather, he had a long history in civil rights. Two years after the Oppenheimer case he represented Arthur Miller in his appearance before a congressional committee. From 1932 to 1945 he had been dean of the University of Wisconsin Law School. He then moved to New York where he joined the distinguished law firm of Paul, Weiss, Wharton, and Garrison. The previous April he had become a trustee of the Institute for Advanced Study, where he had encountered Oppenheimer. I think "encounter" is the right word since, long after it was over, Garrison remarked that though they had been thrown together for six months in a struggle in which he was with Oppenheimer for many hours every day, he never felt he had really gotten to know him.

It was Garrison's intention, once the legal process was started, to turn the representation in the hearings themselves to someone else—an experienced litigator. But he could not find one. Either they were already committed to something else, or they didn't wish to get involved with defending Oppenheimer and his left-wing baggage—another sign of the times. Garrison got permission from his firm to devote himself to the case on a pro bono basis. While Oppenheimer was well to do, paying for this kind of legal help would have ruined him financially as well as emotionally.

The first task that Garrison and his associates faced was drawing up a response to the charges in Nichols's letter. The concern was that this letter might be leaked to the press without the benefit of a response. They asked for an extension for filing this rebuttal and, in the end, they were given until March 4. Meanwhile it became increasingly clear what they were facing. Very early in Garrison's discussions with Strauss, the matter of clearance came up. It was obvious that classified material would be discussed, and Garrison wanted clearance so he could study it and be present when it appeared in testimony. Strauss assured him that the likelihood of this was small. This, of course, was false. In fact, during the hearing a classification expert was on hand at all times to warn witnesses when they needed to move into a secure session.

The AEC had chosen as its attorney a tough prosecutor named Roger Robb, who had been an assistant U.S. attorney. They had obtained for Robb, in eight days, an emergency Q clearance. Once the hearing got under way, Garrison asked for the same, but he was never cleared. This led to the following Orwellian situation. When classified information was discussed, Garrison and the rest of Oppenheimer's attorneys were required to leave the room. But Oppenheimer, whose clearance had been suspended, was allowed to stay. Whether this made any difference in the final result is impossible to say. But other things surely made a difference. In a real jury trial there is a "blank pad" rule. The judge and jury begin the trial with a blank pad, with none of the evidence that is to be presented at trial before them. In this case, the three members of the AEC panel studied such things as Oppenheimer's FBI file and other pertinent documents containing derogatory information—some classified—well before the procedure began. When Garrison asked for the same privilege, it was denied him. Moreover, Garrison supplied a witness list before the hearing began. Robb refused to do this, and his argument was bizarre. He argued that if Garrison and his people knew who the witnesses were going to be, they might try to influence them, especially if they were scientists.

Considering what Robb later did with Edward Teller (to be discussed shortly), the hypocrisy of this argument is blatant.

The first task of the AEC was to choose a panel. This became the responsibility of the AEC general counsel, William Mitchell, though of course Strauss would have the final say. As chairman of the panel, Gordon Gray was selected. He had been assistant secretary of the army when Eisenhower was chief of staff. He was a Democrat, and it is said that Eisenhower played a role in his selection with the idea that choosing a Democrat might blunt criticism. At the time he was chosen, Gray was president of the University of North Carolina. During the hearing, he was the one who consistently delivered the refusal to Garrison's requests for such things as having time to examine some document that Robb had just produced from thin air. Each time Gray delivered one of these refusals, it would be accompanied by a little homily on the fairness of the procedure. The other two members of the panel were Thomas A. Morgan, a businessman and the former president of the Sperry Gyroscope Company, and Ward V. Evans, who had been chairman of the chemistry department at Northwestern University. Morgan did not have much to say during the hearings, but Evans was quite remarkable. When a fellow chemist appeared as a witness, Evans took the opportunity to ask if he knew the whereabouts of various of his students. He also wanted to know if witnesses were Communists or if they had known any Communists. I would imagine that Oppenheimer and his attorneys were amazed when it was Evans who dissented from the majority verdict to deny Oppenheimer his clearance.

The hearing, which began on April 12, was held in Room 2022 of the Atomic Energy Building T-3, in a room that had been transformed into a miniature courtroom. It ended on May 6. An enormous amount has been written about what happened during those three weeks.* Nothing replaces the experience of reading the transcripts. If you are familiar with the issues and the people, they read like a great play. (In fact, in the mid-1960s they were made into a play by the German playwright Heinar Kipphardt. In performance I did not think it was very successful because I knew many of the people being depicted on stage, and there was very little resemblance. Rabi, for example, came across as someone who might be at home managing a delicatessen.) It had not been the AEC's intention to release the transcripts at all. Indeed, Gray made this explicit at the beginning of the hearings. Witnesses testified with the understanding that what they said would be kept secret. But in June, when the verdict of the commission was under appeal, the AEC decided to release the transcripts. People who had been witnesses at the hearing had to be called to obtain their permission. My guess is that by this time the AEC was beginning to feel the heat that was already being generated in the scientific community by the actions of what many people felt had been a kangaroo court. By releasing the transcripts, the commission may have felt it would show that its activity had been fair and honorable. In any event, on June 15 the transcripts were released. They had been printed by the United States Government Printing Office and sold for $2.75. They were 993 pages long, in rather small type. A separate document was later issued which had the details of the commission's final verdict. It was a mere 67 pages long. There was great interest at the time, but by 1971, when the MIT Press reissued the full transcript, they had a very hard time finding a copy to print from. Now it is out of print and difficult to find in any form.

For someone like myself, trying to reconstruct these events, there are several reasons why it is necessary to read the transcripts.

---

*In my view, the best account to date is Stern.

In the first place, the sheer length—700,000 words of testimony—gives some impression of what it must have been like for Oppenheimer. He sat there day after day, chain smoking, when he wasn't testifying, listening to himself either being vilified or defended. It's amazing that he got through it. One also realizes that the witnesses with negative testimony were a relatively small fraction of the whole. Six hundred and fifty-five pages of positive testimony took place before the first witness was called by Robb. This was the Berkeley chemist Wendell Lattimer, who had joined Lawrence in an attempt to activate the crash program to build Teller's super. On cross-examination it was revealed that he knew nothing about the debate among the advisers. Evans asked him if he knew any Communists, and Lattimer gave the remarkable answer, "Yes, I have known Communists. They planted a Communist secretary on me at one time during the war until the FBI discovered her. The army sent her to me. This is the only intimate connection I can recall." Evans chose to let this drop. Bits like this make reading the transcripts essential.

There are also some curious insights into Oppenheimer. He put on record in the transcripts something that I have characterized as a sort of "autobiography." He was examined on it. At one point he writes, "In 1937, my father died; a little later, when I came into an inheritance, I made a will leaving this to the University of California." It is true that six years after his mother died, his father died of a heart attack. During these last years he and Oppenheimer had grown closer together. His father spent time in California and even visited the ranch in the Pecos. But why in this hearing did Oppenheimer bring up the subject of his will? Indeed, one wonders if in 1954 this was still a provision in his will. When he died in 1967, I am not aware of any large benefaction to the University of California.

Then there was the matter of his brother. Garrison questioned him about Frank. Oppenheimer remarked, "My brother had learned to be a very expert flutist. I think he could have been a professional. He decided to study physics. . . . We were quite close, very fond of one another. He was not a very disciplined young man. I guess I was not either. He loved painting. He loved music. He was an expert horseman. We spent most of the time fiddling around with horses and fixing up the ranch." Then he added, "He worked fairly well at physics but he was slow. It took him a long time to get his doctor's degree. He was very much distracted by his other interests." As I said before, it could not have been easy to be Oppenheimer's brother.

At the hearing the two most effective witnesses against Oppenheimer were Edward Teller and Oppenheimer himself. It was not entirely Oppenheimer's fault. Garrison had made a kind of policy decision that governed his own behavior during the hearings. He felt there was not much chance that Oppenheimer would win, and there would be no chance if his lawyers became really aggressive. Unlike a proper judicial process, once the AEC commissioners ratified whatever decision the panel made, there was no further appeal possible unless one went to the president. Garrison was sure that Eisenhower would not involve himself in this. Thus he would protest some outlandish tactic by Robb and when, as was almost inevitably the case, his protest was rejected, he would back off almost apologetically. In particular, Robb had a canonical tactic that worked very effectively against Oppenheimer. He would begin a line of questioning about something, such as the interview that Oppenheimer had had with Pash and Johnson in Berkeley a decade earlier—an interview the details of which Robb had at hand and Oppenheimer had forgotten. He would then lead Oppenheimer down some path and, just at the right moment, produce the transcript of the hearings, or whatever, to show that Oppenheimer had mischaracterized what had happened. This seemed completely to disconcert Oppenheimer, who would occasionally ask Robb to tell him what actually had happened. Neither Oppenheimer nor his lawyers would be furnished copies of this material beforehand. Needless to say, this would not have been allowed in any court of law, but it contributed to the negative impression Oppenheimer made on the panel. As Oppenheimer once told me on one of our train rides, at the time he felt that the whole thing must be happening to someone else. Teller was another story.

The most complete, and certainly the most interesting, of the various versions that Teller has given over the years of the reasons for his testimony, is to be found in his memoirs. A whole chapter is devoted to it. Since he died on September 9, 2003, this is presumably his last word. His basic contention is that he did not think Oppenheimer's opposition to the hydrogen bomb was sufficient to deprive him of his clearance. He writes, "I was certain that Oppenheimer would be cleared. I thought the only good that could come out of the affair would be that the testimony might demonstrate the degree to which Oppenheimer's advice was wrong-headed. I went over all the incidents where I felt his advice had worked against national well-being. . . ."

Then he goes on,

"Earlier that year I had seen Oppenheimer at a physics conference on the East Coast [the Rochester Conference discussed earlier]. He asked me whether I believed he had done anything 'sinister.' I assured him I did not. He then asked me if I would talk with his lawyer, Lloyd Garrison, and I agreed. Eventually, I did see Garrison and the co-counsel Herbert Marks. I don't recall just when our meeting took place. At the time of the interview, I was only vaguely aware of the charges; some were related to Oppenheimer's opposition to developing the hydrogen bomb and some were related to his and his family's involvement with communism. I was unaware of the complexity of the charges related to Chevalier—that when Groves had pressured Oppenheimer to name the spy that Oppenheimer had mentioned to security officers, Oppenheimer had given the name of his friend Chevalier."

Teller continues,

"As soon as I [was] seated in the office, Oppenheimer's lawyers asked me whether I was familiar with the accusations against Oppenheimer. Because I felt that the interview was a waste of time given my personal conviction Oppenheimer had done nothing sinister, I somewhat inaccurately told them that I was familiar with the charges. Consequently they did not mention the Chevalier incident. Rather, they devoted all their efforts to convincing me, indirectly, that Oppenheimer's actions in regard to the hydrogen bomb were innocent of disloyalty. In consonance with them, I had already made up my mind about this. For the half hour or more I was with them, they described the magnificent contribution that Oppenheimer had made in wartime Los Alamos and emphasized that Oppenheimer's work had demonstrated his great dedication to his country. They mentioned nothing I didn't already know and believe. . . . I left unimpressed with their comments and without having changed my mind: I would testify that Oppenheimer was a loyal citizen."

Leaving aside the casual characterization of Chevalier as a "spy," one is struck by how ingeniously Teller evades the real issue. Would he testify that Oppenheimer was not a security risk, and therefore that his clearance should be restored? Many years later Garrison described the same meeting. Here is what he wrote, "At my request he [Teller] came to my law office, not very long before the hearings were to begin. He was reluctant to come, and insisted on seeing me alone. I asked him about his associations with Robert and his opinion of Robert's loyalty. His feelings towards Robert were not warm, but he did not challenge his loyalty. He expressed lack of confidence in Robert's wisdom and judgment and for that reason felt that the government would be better off without him. His feelings on this subject and his dislike of Robert were so intense that I finally concluded not to call him as a witness. I do not recall his indicating to me that he would be seeing Mr. Robb, who was later to call him as a witness. In any event, when Dr. Teller did take the stand, his testimony did not depart in any substantial respect from what he had said to me in our interview."

I will shortly describe Teller's encounter with "Mr. Robb," but first here is Hans Bethe on the same subject. He and his wife had gone to see Teller in Washington shortly before he was scheduled to testify. "I tried to persuade him to testify in favor of Oppie—or, at least, not against him. Most of Teller's arguments against Oppie involved matters of judgment on which the two of them had differed, and were, in my view, no basis for conducting a security hearing. But Teller was immovable."

In his memoirs, Teller informs us that when he arrived early on April 28 to testify, Robb came into the room where he was waiting and asked, "Should Oppenheimer be cleared?" To which Teller says he replied, "Yes, Oppenheimer should be cleared." At this point Robb gave Teller, he recalls, a short bit of previous testimony to read. Even Teller was somewhat surprised by this totally unethical and probably illegal act. Remember, at this point the witnesses had been guaranteed that their testimonies were sealed. As Teller writes, "At that time, the proceedings of the hearing were secret. For a moment I thought of saying no to Robb. But then I remembered that I had listened to Oppenheimer's lawyers for more than half an hour. Having given them that much time, I thought it would be fair to read the material that Robb wanted me to see." The equation Teller proposes here between a half-hour spent with Oppenheimer's lawyers and the time spent to read secret, privileged testimony taken from the hearing defies comprehension. Can Teller really believe this? And Robb, what is he?

The testimony that Robb gave Teller was the portion of his cross-examination of Oppenheimer about the Chevalier incident in which Oppenheimer says that he did what he did because he was an "idiot." This shocked Teller, he tells us, to such an extent that he was now prepared to change his mind about Oppenheimer's security clearance. Here is the pertinent part of the testimony that resulted. "Q" is Robb and "A" is Teller.

Q. To simplify the issues here, perhaps, let me ask you this question: Is it your intention in anything that you are about to testify to, to suggest, that Dr. Oppenheimer is disloyal to the United States?

A. I do not want to suggest anything of the kind. I know Oppenheimer as an intellectually most alert and a very complicated person, and I think it would be presumptuous and wrong on my part if I would in any way analyze his motives. But I have always assumed, and now assume that he is loyal to the United States. I believe this, and I shall believe it until I see very conclusive proof to the opposite.

Q. Now, a question which is a corollary of that. Do you or do you not believe that Dr. Oppenheimer is a security risk?

A. In a great number of cases I have seen Dr. Oppenheimer act—I understood that Dr. Oppenheimer acted in a way which for me was exceedingly hard to understand. I thoroughly disagreed with him in numerous issues, and his actions, frankly appeared to me confused and complicated. To this extent, I feel that I would like to see the vital interests of this country in hands which I understand better and therefore trust more.

Commenting on this in his memoirs, Teller notes, "But my most glaring error was that I did not reveal that Robb had shown me Oppenheimer's testimony about Chevalier. As a result, everyone assumed that my testimony was meant to accuse Oppenheimer for his opposition to the H-bomb. My comments were only meant to question his behavior in regard to Chevalier, but my statement could have referred with equal ease to Oppenheimer's opposition to the hydrogen bomb. By using general terms I made it impossible for anyone to distinguish what I meant to criticize. By talking about Oppenheimer's behavior as a friend I had produced the impression that I was talking about Oppenheimer in regard to the hydrogen bomb."

Can anyone reading Teller's testimony really believe this? That it is an invention is confirmed a little later in the testimony. This time it is Gray who is asking the questions. He begins by explaining the function of the panel:

MR. GRAY. . . . We are asked to make a finding in the alternative, that it will not endanger the common defense and security to grant security clearance to Dr. Oppenheimer.

I believe you testified earlier when Mr. Robb was putting questions to you that because of your knowledge of the whole situation and by reason of many factors about which you have testified in very considerable detail, you would feel safer if the security of the country were in other hands.

THE WITNESS. Right.

MR. GRAY. That is substantially what you said?

THE WITNESS. Yes.

MR. GRAY. I think you have explained why you feel that way. I would like to ask you this question: Do you feel that it would endanger the common defense and security to grant clearance to Dr. Oppenheimer?

THE WITNESS. I believe, and that is merely a question of belief and there is no expertness, no real information behind it, that Dr. Oppenheimer's character is such that he would not knowingly and willingly do anything that is designed to endanger the safety of this country. To that extent, I would say that I do not see any reason to deny clearance.

If it is a question of wisdom and judgment, as demonstrated by actions since 1945, then I would say one would be wiser not to grant clearance. I must say that I am myself a little bit confused on this issue, particularly as it refers to a person of Oppenheimer's prestige and influence. May I limit myself to these comments?

There were two reasons why this testimony was so devastating. The first is the source. By this time Teller had replaced Oppenheimer as the icon of nuclear weapons. His views were listened to with deference, certainly by the panel. The second was the message. Teller was offering a way of splitting the baby that the majority of the panel found attractive. One could proclaim Oppenheimer's loyalty but at the same time deny his clearance. Just think what it would have meant if the panel had found Oppenheimer disloyal—a traitor. What would one do then? Try him for treason? A clue to how they felt about this is what happened when Borden was brought on as Robb's last witness. The panel signaled that they were not interested in his charges that Oppenheimer was a spy, and Garrison did not even bother to cross-examine him. When the majority delivered its opinion on May 27 they noted in their recommendation, ". . . Of course, the most serious finding which this board could make as a result of these proceedings would be that of disloyalty on the part of Oppenheimer to his country. For that reason we have given particular attention to the question of loyalty, and we have come to a clear conclusion, which should be reassuring to the people of this country, that he is a loyal citizen. If this were the only consideration, therefore, we would recommend that the reinstatement of his clearance would not be a danger to the common defense and security."

But then they go on to say, "We have, however, been unable to arrive at the conclusion that it would be consistent with the security interests of the United States to reinstate Dr. Oppenheimer's clearance, therefore, do not so recommend.

"The following considerations have been controlling in leading us to our conclusion:

"1. We find that Dr. Oppenheimer's continuing conduct and associations have reflected a serious disregard for the requirements of the security system.

"2. We have found a susceptibility to influence which could have serious implications for the security interests of the country.

"3. We find his conduct in the hydrogen bomb program sufficiently disturbing as to raise a doubt as to whether his future participation, if characterized by the same attitudes in a Government program relating to the national defense, would be clearly consistent with the best interests of security.

"4. We have regretfully concluded that Dr. Oppenheimer has been less than candid in several instances in his testimony before this Board."

It was Ward Evans, with all his bizarre questions about recognizing Communists, who picked up on the paradox—loyal but a security risk. He wrote a brief dissent in which he noted,

"Most of the derogatory information was in the hands of the Committee when Dr. Oppenheimer was cleared in 1947. They apparently were aware of his associations and his left-wing policies: yet they cleared him. They took a chance because of his special talents and he continued to do a good job. Now when the job is done, we are asked to investigate him for practically the same derogatory information. He did his job in a thorough and painstaking manner. There is not the slightest vestige of information before this Board that would indicate that Dr. Oppenheimer is not a loyal citizen of his country. He hates Russia. He had communistic friends, it is true. He still has some. However, the evidence indicates that he has fewer of them than he had in 1947. He is not as naive as he was then. He has more judgment; no one on the Board doubts his loyalty—even the witnesses adverse to him admit that— and he is certainly less a security risk than he was in 1947, when he was cleared. To deny him clearance now for what he was cleared for in 1947, when we must know he is less of a security risk now then he was then, seems hardly the procedure to be adopted in a free country."

During Teller's testimony, one exchange with Robb had a special resonance for me. It went as follows—with "Q," as usual, being Robb.

Q. Doctor, the second laboratory, is that the one in which you are now working at Livermore?

A. That is the one at which I have been working for a year and at which I am now working part time. I am spending about half my time at the University of California in teaching and research and half my time in Livermore.

Q. Did you have any difficulty recruiting personnel for that laboratory?

A. Yes, but not terribly difficult.

Q. Did you get the personnel you needed?

A. This is a question I cannot really answer, because it is always possible to get better personnel. But I am very happy about the people whom we did get and we are still looking for very excellent people if we can get them, and I am going to spend the next 3 days in the Physical Society in trying to persuade additional young people to join in.

As it happened, I was one of the "additional young people" that Teller spoke to. I was still a year away from finishing my thesis, but a Harvard professor who had spent a sabbatical at Berkeley and Livermore had given my name to Teller. I had only a vague idea that Livermore was a weapons laboratory and, in any case, I was not really looking for a job. But I was curious to meet Teller, and since I was going to the Physical Society meeting in Washington anyway, I made an appointment. I recall meeting him in the lobby of the hotel where we were all staying and in which the lectures and the rest took place. He suggested going to his suite, which would be quieter. He said he was going to lecture shortly on some meson theory he was working on, and the way he prepared was to give a preview to someone like myself, which he proceeded to do while pacing up and down the living room of his suite. He told me to interrupt him with questions. In truth, what I understood of his theory did not seem very interesting, so I had no questions. At one point he stopped pacing and said cryptically that it was much more pleasant to talk physics than politics. I now wish that I had asked him what he meant. I guess I did not make much of an impression because I never heard from Livermore.

Following the panel's verdict, the next step in the process should have been routine. But nothing involving Oppenheimer was ever routine. General Nichols, as general manager of the AEC, had the responsibility to convey the results of the Gray panel as a "recommendation" to the five AEC commissioners who were to pass final judgment. Garrison thought so little of this that he did not even insist on seeing Nichols's "recommendation." He should have known better—not that it would have helped in the final analysis. In addition to confirming the judgment of the majority, Nichols added a charge of his own: perjury. He wrote,

"In my opinion, Dr. Oppenheimer's behavior in connection with the Chevalier incident shows that he is not reliable or trustworthy; his own testimony shows that he was guilty of deliberate misrepresentations and falsifications either in his interview with Colonel Pash or in his testimony before the Board; and such misrepresentations and falsifications constituted criminal, *** dishonest *** conduct."

Garrison had no chance to respond to this because he was never shown the document. In any event, it would not have mattered. Strauss, who had coached Robb during the hearings, was merely waiting to write his final verdict on Oppenheimer. He wrote the majority report for the AEC, which was issued on June 29. It is brimming with vitriol. The sole dissent was written by the Princeton physicist Henry DeWolf Smyth. He later recalled that he had stayed up all night writing, even though he didn't much like Oppenheimer. He liked even less what had happened.

With this, Oppenheimer's government service was over. For a while he wondered if he would be able to keep his position as director of the Institute for Advanced Study. On February 15, in view of the charges pending against him, Oppenheimer had offered his resignation. The trustees of the Institute held a special meeting in which they reaffirmed their complete confidence. There were two absentees. One of them was Strauss.

Some idea of what the hearing meant to Oppenheimer and his family was found on something his son Peter, then a teenager, wrote on the blackboard he had in his room in Princeton. "The American Government is unfair to Acuse [sic] Certain People that I know of being unfair to them. Since this is true, I think that Certain People, and may I say, only Certain People in the U.S. government, should go to HELL.

Yours truly
Certain People "

Oppenheimer receiving congratulations from Edward Teller
upon winning the Fermi Award, 1963

THE INSTITUTE

Members are kindly requested to play touch football out of earshot of the library.

—Robert Oppenheimer

THE FIRST TIME I actually met Oppenheimer was in the spring of 1957. He had come to Harvard to deliver the first of six public lectures which he called "The Hope of Order."* It was an occasion. At Sanders Theater, the largest lecture venue on the campus, its twelve hundred seats were filled and another eight hundred people could listen on speakers in the so-called New Lecture Hall. The lecture attracted not only the university community but peo-

---

*This was a William James Lecture. They were given alternately by the philosophy and psychology departments. This time, as a gesture of support to Oppenheimer, the two departments had united to make their choice. (I am grateful to Morton White, now of the Institute for Advanced Study and then of the Harvard philosophy department, for explaining this to me.) In the audience was William James, Jr., the philosopher's son.

ple from all over Boston. Seated in front of me were two of those wonderfully elegant ancient Boston ladies with blue hair. I once asked Rabi if his motivation for doing science had dimmed after he won the Nobel Prize. He answered by saying that when one of these Brahmin ladies had been asked if she traveled, she replied, "Why should I go anywhere when I am already there?" At one point in his talk, Oppenheimer said he found it necessary to write down a mathematical expression. As the ladies clutched each other for reassurance, he wrote on a blackboard the expression $[x, p]=i\hbar$, the canonical commutation relation in quantum mechanics between position and momentum, from which Heisenberg's uncertainty principle is derived. I wanted to assure them that it was just Oppenheimer showing off.

Of course the physics department was out in force. At the time, since there was not much mention of it, I did not fully realize how many members of the department had been involved either with Los Alamos or with Oppenheimer in California. The chairman of the department, Kenneth Bainbridge, had been in charge of preparing the test site at Trinity. After the explosion he had said to Oppenheimer, ". . . Now we are all sons of bitches"—a reflection that was considerably less baroque than Oppenheimer's reference to the Gita. Norman Ramsey, a professor in the department, had been on the team on the island of Tinian that had assembled Fat Man, the bomb that was dropped on Nagasaki. I had taken over Ramsey's course on nuclear physics when he had been called to Washington to defend Wendell Furry in a television debate. In the mid-1930s Furry had been at Berkeley with Oppenheimer; they had published some fundamental papers on quantum electrodynamics. I had taken introductory physics with Furry. Roy Glauber, who was a young assistant professor, had been at Los Alamos before he had gotten his Ph.D., along with Theodore Hall, who went to Harvard at the age of sixteen and Los Alamos at eighteen, the youngest person on the technical staff.

What was not known for certain until the 1990s was that Hall had spied for the Russians. The information he gave them about the design of the implosion weapon was probably as important as what Klaus Fuchs had conveyed. It confirmed Fuchs's information and helped to convince Stalin that the whole thing was not a hoax. The presence of these totally undetected espionage agents shows once again how absurd the Oppenheimer hearings, with their emphasis on the minutiae of how often Oppenheimer had dinner with Chevalier and the like, were.

At the lecture was, of course, Schwinger, who was the reigning theoretical genius in the department. Percy Bridgman, Oppenheimer's beloved physics professor, was still alive, but I do not know if he was at the lecture. I would imagine that Kemble was there. I was there because I was still at Harvard, even after getting my Ph.D. in 1955.

I had become the house theorist for the Harvard cyclotron. It was a two-year appointment. When I started, I was told that my main job would be helping the experimenters with the theory of their experiments. It soon became clear that these particular experimenters knew a great deal more about the theory than I did. If anything, they could help me. This left me free to do my own work. I had written my dissertation on some electromagnetic properties of the nucleus of heavy hydrogen—the deuteron—and this is what I continued to work at. But by 1957 I was looking for a job. Without any great expectation of being successful—I knew it was terribly competitive—I had applied to the Institute for Advanced Study for a fellowship. I saw the Institute as a kind of Camelot, with Einstein—who had died two years earlier—the reigning king. In fact I had even spent a couple of days there visiting a friend and had actually seen Einstein on one of his daily walks. I knew that Oppenheimer was the director and that he too was legendary. I asked Schwinger for a letter of recommendation and, when he complained about the number of such letters he had to write, I told him that if he ever needed one from me I would be only too happy to oblige. Victor Weisskopf, who had been a group leader at Los Alamos and was now at MIT, also wrote a recommendation. Nonetheless I was truly amazed—and absolutely thrilled—when I received a letter of acceptance. The annual stipend of $4,500 seemed like a small fortune. I immediately went to Schwinger and thanked him. I told him that while I was at the Institute I intended to keep my mouth shut. He said that was probably a good way of staying out of trouble.

When I was reviewing the corpus of Oppenheimer's work in physics for this book, I noticed that some of his papers of the 1930s also dealt with the electromagnetic properties of the deuteron. Perhaps that is why I was admitted. In any case, not long after his letter arrived, there was Oppenheimer giving a lecture at Harvard. Nothing that has been written about his charisma as a public lecturer has been exaggerated. It was a mixture of phrasing that was both elegant and somewhat obscure. You were not quite sure what he meant, but you were sure that it was profound and that it was your fault that you didn't see why. He spoke in a curious clipped way—an accent that was not quite British but not exactly American either. T. S. Eliot, who incidentally had been at the Institute a few years earlier, spoke with a similar accent.

After the lecture I decided to go onto the stage and introduce myself to Oppenheimer. I was, after all, going to be one of his charges in a few months. I went up to him, and he looked at me with what I distinctly remember as icy hostility—his students referred to it as the "blue glare." It was clear that I had better explain—quickly—why I was bothering him. When I told him I was coming to the Institute that fall, his demeanor completely changed. It was like a sunrise. He told me who would be there— an incredible list. He ended by saying that Lee and Yang were going to be there and that they would teach us about parity. (This was a sensational discovery that had just been made about the breakdown in symmetry between left- and right-handed descriptions in physics. T. D. Lee and C. N. Yang had been at the center of it and, indeed, in the fall of 1957, my first semester at the Institute, they won the Nobel Prize.) Then Oppenheimer said, with a broad smile, "We're going to have a ball!" I will never forget that. It made it clear to me why he had been such a fantastic director at Los Alamos.

The Institute that I found in 1957 was radically different from the one that Oppenheimer took over in 1947. To understand this it is useful to recapitulate some of its history. It began in 1929 when Louis Bamberger and his sister Julie Carrie Fuld were looking for an educational enterprise they could create using the money they had made from their New Jersey department store. Indeed, they had just sold it to R. H. Macy & Company for eleven million dollars in cash plus Macy's stock, which was soon devalued in the stock market crash. Their first idea was to found a medical school, which they thought might help offset the problems that potential Jewish doctors were having getting into such schools. To guide them they consulted Abraham Flexner, a recognized authority on medical education. Flexner thought that opening a new medical school so close to New York, where there were already such schools and the hospitals necessary to serve them, would be a mistake. Instead he proposed that they found an entirely different sort of institution, a place in which there would be no undergraduate students but which would have a brilliant faculty with basically no other obligation except to do its scholarly work. A model he had in mind was All Souls College at Oxford. This appealed to Bamberger and his sister. Flexner then set out to generate plans.

By 1930, they were beginning to take concrete form. His notion was that there should be "schools"—departments—in fields like mathematics, economics, and history. Physics was less attractive because it seemed to be done well at other institutions. A field like mathematics was chosen because it was thought to be fundamental, and the infrastructure would be inexpensive—chalk and blackboards—and because there might be agreement as to which mathematicians were at the top of their profession—unlike, say, professors of English. When the *New York Times* reported on these plans, it noted an initial endowment of five million dollars. It also remarked that the Institute would have no athletic teams. (I found this particularly amusing. When I arrived in Princeton in the fall of 1957, it was at the beginning of the football season. There were signs of all sorts hanging from the various Princeton eating clubs, exhorting the football team to beat whatever Ivy League school was on the schedule that weekend. I tried to promote the idea of having a sign made that would read "Beat Copenhagen"—a reference to the Bohr institute—which I planned to display on Fuld Hall at the Institute, but I got no takers.)

Then Flexner set about to find a faculty that met his requirements. What happened next transformed the Institute. Flexner had gone to Pasadena in January 1932 to consult with the physicist Robert Millikan, who was, as its president, in the process of building up Cal Tech's physics department. Oppenheimer was coming there for a few months a year as part of this. But, as it happened, Einstein was also at Cal Tech for the winter semester. Never one to miss an opportunity, Flexner set out to recruit Einstein, who was in the process of leaving Germany. This required following him to

Oxford where Einstein consented, with the appointment starting in October 1932. Einstein was, of course, a physicist—but he was also Einstein.

Now the other appointments began to fall into place. They included the mathematicians John von Neumann and Hermann Weyl. Von Neumann was considered to be the most brilliant of the young mathematicians and Weyl one of the old masters. In addition there were people like Erwin Panofsky, the great art historian, as well as the logician Kurt Gödel. All the examples I have mentioned were refugees from Hitler's Europe, but there were Americans as well. It was a superb faculty, just as Flexner had proposed.

The Institute at the time did not have a campus of its own but used facilities around Princeton. In particular, the mathematicians and physicists had offices in Fine Hall, where the Princeton mathematicians also had their offices. When Einstein entered that building for the first time after his appointment, he encountered something familiar. In 1921 a physicist named Dayton Miller had claimed to have done an experiment which, if true, would have destroyed the theory of relativity. Einstein was lecturing in Princeton then and was told about it. He said in German, *"Raffiniert ist der Herr Gott, aber boshaft ist er nicht"*—God is subtle but not malicious. He was asked his permission to have this carved—in German—over a fireplace in Fine Hall, which happened in 1930. The last time I looked, it was still there.

By 1935 a site had been selected a few miles east of Princeton—the Olden Farm—where the Institute's facilities would be built. The main house on the farm became the director's home. (This is where Oppenheimer had the van Gogh when I saw it in 1957.) The site was, incidentally, less than a mile from Einstein's house on Mercer Street, so he could walk back and forth. In the fall of 1938, Flexner announced plans to construct the first of the Institute's buildings, Fuld Hall. A 1940 photograph shows it looking as it did in 1957. It still looks the same, but there are number of new buildings that make the campus, at least to me, almost unrecognizable. Einstein had his office in Fuld Hall, and on the top floor was a cafeteria. On the bottom floor there was, and still is, a tea and newspaper reading lounge. Afternoon teas at the Institute were always a high point of the day. Until the fall of 1957 there were only some little grey shacks on the campus available for faculty housing. One of the inhabitants, Morton White, the philosopher and intellectual historian, told me he had to shovel coal for the furnace when he spent the 1953–1954 academic year there. In 1969 he became a permanent member. In his autobiography, A *Philosopher's Story*, White tells of a dinner he hosted in 1953 at which Oppenheimer and Kitty were guests. The conversation turned to Gödel's theorem in logic which demonstrates that there are true but necessarily unprovable statements in mathematics. White remarked that Oppenheimer was wrong about what he was claiming about the contents of the theorem, at which Kitty remarked, "'Robert is never wrong!' and Robert did nothing to disown her foolish remark, probably because he feared what the effect of doing so would be when they got home." By the fall of 1957 a set of newly constructed apartments designed by Marcel Breuer was opened. I occupied one of them. At the time I thought it was the nicest place I had ever lived in. There was no trace of the coal furnaces. Oppenheimer had selected the furniture. As I recall, some of it was from Knoll and some from the Design Workshop in Cambridge. I visited one of the apartments in the spring of 2003, and the furniture looked familiar.

In 1936, Flexner began to groom his successor, a man called Frank Aydelotte, who had been one of the original Institute trustees and was president of Swarthmore College. The ambiguity of this role, coupled with the natural combativeness of university professors, led to several years of internecine warfare before Flexner actually departed in 1940, at least formally. He still lurked in the background. After all, it was *his* Institute. Aydelotte introduced the idea of temporary memberships for young postdoctorals. Before this, visitors had usually been senior people. Both Rabi and Dirac had spent time at the Institute. But there were no professorial appointments in physics—or, indeed, anything else—during Aydelotte's tenure, though Pauli spent the war years there as a visitor. In 1944, Aydelotte announced that he would retire when he was sixty-five, which would have been in 1946. For some reason this outraged Flexner, who felt that Aydelotte had no right to desert his post. A compromise was reached: Aydelotte would stay until he was sixty-seven, and a search committee would look for the next director.

By August 1945, Oppenheimer was making plans for what he wanted to do after he left Los Alamos, which would be soon. His old job at Berkeley was open to him, but there were reasons why he was beginning to think he did not want to return. On the one hand, there were strained relations with the physics department's longtime chairman Raymond Birge. As early as 1943 he had tried unsuccessfully to persuade Birge to make an offer to Richard Feynman. In a letter to Birge he quoted something that the Princeton physicist Eugene Wigner had said about Feynman: "He is a second Dirac, only this time human." (It should be noted that Wigner was Dirac's brother-in-law.*) Birge refused to consider the appointment, and after the war Feynman went to Cornell. It is also clear that relations with Lawrence were not good. Lawrence paid a visit

---

*Dirac always introduced his wife as Wigner's sister, as in "I would like you to meet Wigner's sister."

to "Y"—the code name for "Site Y," which was Los Alamos—in the summer of 1945. He and Oppenheimer had a quarrel about something to do with Berkeley. On August 30, Oppenheimer wrote to Lawrence from Perro Caliente. It is a remarkable letter, free of the baroque tone of so many of Oppenheimer's letters. In part he writes,

". . . We have been at the ranch some days now, and I'm beginning to recover a little of the sanity that had all but vanished by the weekend you visited Y. I have very mixed and sad feelings about our discussions on Berkeley. I meant them in a far more friendly, tentative and considerate spirit than they appeared to you: and was aware and tried to make you aware at the time that fatigue and confusion gave them a false emphasis and color. It may seem odd and wrong to you that the lack of sympathy between us at Y and the California administration* over the operation of the project could make me consider not coming back: I think it would not have seemed so odd if you had lived through the history as we did, nor so hard to understand if you remembered how much of more of an underdogger I have always been than you. This is a part of me that is unlikely to change, for I am not ashamed of it; it is responsible for such differences as we have had in the past, I think; I should have thought that after the long years it would not be new to you. In any case it seemed a little more honorable to tell you of my misgivings, however unfortunate the timing. My own thoughts and plans have not become clear, though they no doubt will. But it must be apparent that your very strong, very negative reactions would, if confirmed, and quite apart from the view of others involved, tend to carry a considerable weight with me, since any

---

*This is presumably a reference to tensions between Los Alamos and the University of California, which formally administered the project.

fruitful future in Berkeley would have to depend, not on identity certainly, but on a certain mutual respect for non identical points of view. . . ."

One might naively think that the feelings that prompted this letter would have made it impossible for Oppenheimer to return to Berkeley. But academic courtships are like most others—they are not over until they are over. Birge and Lawrence weighed in with very conciliatory communications and, in the end, Oppenheimer accepted an arrangement that was very similar to what he had had in the 1930s, with its center of gravity at Berkeley and regular visitations at Cal Tech. Meanwhile the Institute search committee, now chaired by Lewis Strauss, was actively looking for a new director. Strauss himself was briefly in the running. In early 1947, he was authorized to approach Oppenheimer. At the 1954 hearing Oppenheimer was asked by his counsel, Lloyd Garrison, to describe the circumstances of his coming to the Institute. He explained,

"I came in the late summer, I think, of 1947. I had been a professor at California Institute of Technology and at the University of California at Berkeley. In late 1946 or early 1947, the present Chairman of the Atomic Energy Commission [Strauss] was chairman of the nominating committee to seek a new director to succeed Dr. Aydelotte at the Institute and he offered me the job stating that the trustees and the faculty desired this.

"I did not accept at once. I like California very much, and my job there, but I had, as will appear, not spent very much time in California. [Most of his time after the war ended was being spent in Washington, advising the government.] Also, the opportunity to be in a small center of scholarship across the board was very attractive to me. Before I accepted the job, and a number of conversations took place, I told Mr. Strauss there was derogatory information about me. In the course of the confirmation hearings on Mr. Lilienthal especially, and the rest of the Commissioners, I believe Mr. Hoover sent my file to the Commission, and Mr. Strauss told me that he had examined it very carefully. I asked him whether this seemed to him in any way an argument against my accepting this job, and he said no, on the contrary—anyway, no—in April I heard over the radio I had accepted, and decided that was a good idea. I have been there since."

On March 8, 1947, Hoover sent a file to the Atomic Energy Commission chairman David Lilienthal with information "relative to Julius Robert Oppenheimer . . . and his brother Frank Friedman Oppenheimer." This was the file that contained a summary of the derogatory information the FBI had compiled, and was reviewed by Lilienthal and the other commissioners. Whether this is what Oppenheimer is referring to here, or whether he is conflating various files, I do not know. In any event, the essential thing is that Strauss was fully aware of Oppenheimer's past when he offered the Institute job.

Physics at the Institute when Oppenheimer arrived was one of its weak spots. Einstein was now totally out of the mainstream, and Pauli had chosen to go back to Zurich. There were no other professors in physics, unless one counted von Neumann, who could do anything. Oppenheimer set out to repeat what he had done at Berkeley in the 1930s—to create a school. One of his first and most successful recruits was Freeman Dyson. Dyson once described for me his first introduction to offering a seminar at the Institute. To understand the circumstances one must recapitulate one of the most important developments in postwar physics—quantum electrodynamics.

When the quantum theory was created in the 1920s, it was understood that it would have to be reconciled with the classical theories of electricity and magnetism which, in their domains of application, were brilliantly successful. This work was begun at once, and in the 1930s Oppenheimer and his associates made several contributions to it. But there was a roadblock. The quantum theory kept giving infinite answers for things that had to be finite and well defined. Here is where matters stood when the physicists went off to war. One of the things they learned from their work on radar was how to use microwaves. After the war these techniques were applied to very subtle measurements of properties of the electron, and it soon became clear that there were anomalies which needed quantum electrodynamics to resolve. There then came, in 1947, two apparently different developments, Feynman's and Schwinger's. (There was actually a third, due to a Japanese physicist named Sinitiro Tomonaga. It was remarkable because he had done it in total isolation, which is why it became known only later.)

The Feynman and Schwinger creations were as different as their respective personalities. Feynman's was largely intuitive and pictorial—Feynman diagrams—which made it very accessible. The problem with it, at least for people like me, was that I couldn't see why it worked. There were these wonderful rules. If you followed them you could happily calculate almost anything. But what justified the rules? Schwinger, on the other hand, started from first principles and went forward like an advancing tank. The problem was that to get numbers out of the theory to compare with experimental results, you had to do horrendous calculations. Oppenheimer once remarked that most people present a problem to show how it can be done, while Schwinger presents a problem to show that only *he* can do it. I was at Harvard while all of this was unfolding and could observe several of my colleagues close to nervous breakdowns when they tried to carry out some calculation that Schwinger had assigned them for a thesis.

Then came Freeman Dyson. During the war he had been doing operations research for the RAF and had already published some significant papers in mathematics. In fact he was regarded as one of the best mathematicians in England. He had attended Cambridge, where he had taken a course in quantum mechanics from Dirac. Dyson told me that Dirac simply read from his great book on the subject. When anyone objected, Dirac would note that he had given a good deal of thought as to how to present the theory and this had gone into the writing of the book. As far as he was concerned, this was the only way to present it. By this time Dyson had decided he wanted to do physics, so it was suggested that he go to Cornell to work with Bethe. As an unexpected bonus, there was Feynman.

Dyson was able to watch Feynman at close hand. He concluded that he was half genius and half buffoon. He was also able to study Schwinger's papers. In the course of this he had an epiphany: the two theories were identical! You could have the Feynman rules and the Schwinger first principles at one and the same time. He wrote a series of papers on this which for the first time made the whole business comprehensible to someone like me.

This is what Dyson was going to talk about at the Institute, where he was a visiting member, at a time before his papers had appeared and been digested. He got the treatment. He was scheduled to give five lectures, but in the first two Oppenheimer would not let him talk. He reduced him to silence. I used to see this occasionally during seminars when I was there. You had to be pretty self-confident to deal with it. One person who did was the late Res Jost, a Swiss physicist of very great power and personality. The dialogue went as follows:

OPPENHEIMER: Res, can you explain so and so?

JOST: Yes.

Jost then promptly proceeded with his talk, only to be interrupted by Oppenheimer again.

OPPENHEIMER: I meant, will you explain so and so?

JOST: No.

OPPENHEIMER: Why won't you explain so and so?

JOST: Because you will not understand my explanation, and
you will ask more questions and use up my whole hour.

Remarkably, Oppenheimer seemed to accept this graciously.

Dyson was still nominally a graduate student—he never did
bother to get his Ph.D.—and far too polite to do anything like this.
But Hans Bethe, who understood that what Dyson was trying to explain was very important, appeared at the Institute to give a lecture
that preceded Dyson's third effort. Not only did he speak to Oppenheimer, but in the course of his own lecture he said that no
doubt some of his points had already been made by Dyson. This
worked, and Oppenheimer remained reasonably silent. After the
fifth lecture Dyson found a note in his mailbox that said, "Nolo
contendere. R.O." He was then offered a long-term visitor's appointment at the Institute and in 1953 a professorship. I have an ineluctable memory of going, shortly after my arrival in 1957, to a
party given by some Princeton social figure. She asked Dyson what
he thought of Princeton, and without batting an eye he said,
"Where every prospect pleases and only man is vile."*

By the time I arrived at the institute, the physics faculty was
very strong. There were Dyson, Abraham Pais, and Yang, and they

---

*Some time ago I tracked down the provenance of this quotation, using every-
thing from the *New Yorker* checkers to the Princeton English department. It
comes from a rather racist missionary hymn, "From Greenland's Icy Mountains,"
written by Reginald Heber (1783–1826). I would imagine that Dyson must have
sung this on Sundays in his public school.

would be joined shortly by Lee. Several people had long-term appointments that allowed them to visit whenever they wanted, including Bohr and Dirac. There was a new appointment for the first time in astrophysics—Bengt Strömgren, whom, for reasons I will explain, I got to know better than I expected. Many of the best young physicists had spent time at the Institute. It was the "in" thing to do. Some had come back to Cambridge with "war stories" about how competitive it was. I was told that Oppenheimer had periodic "confessionals" in which you were asked to describe what you had been doing. It was better, they said, to say nothing than to present something that was either trivial or wrong. One Harvard Ph.D. had cracked under the strain. He simply walked out and joined a monastery.

In my own case, I had spent the summer before I went to Princeton at Los Alamos. It was still pretty much a military institution—carefully guarded—and devoted to producing ever more devastating weapons for the cold war. We bachelors lived in barracks that I am sure were there during the war. Some of the senior people were also left over from the war. I had a Q clearance—the one that Oppenheimer had lost—which entitled me to know anything on a "need to know" basis. Since there was nothing I needed to know, no one told me anything. A Harvard colleague who had gotten his degree at about the same time I did was there, and we collaborated on a problem in pure physics while weapons work was going on all around us.

Sometime toward the end of my stay I learned that it would be possible to watch some above-ground atomic bomb testing at the Nevada test site if one paid one's own way, which I was more than willing to do. The bombs that were being tested were candidates for the primaries in hydrogen bombs. At one point on my visit I was given a plutonium "pit" to hold. It was warm to the touch. But nothing prepared me for the sight of the actual explosions—an awesome, vulgar, and terrible beauty. This was the time when it was thought that soldiers might have to fight on battlefields where atomic bombs might be going off. While I was at the site, details of soldiers were being marched to ground zero just after each test to accustom them to these conditions. This experience was what was on my mind when I drove east to Princeton.

At the time I had a Morris Minor convertible with a leaky top that let in air whether or not you wanted the ventilation. By the time I arrived at the Institute I was in a state beyond good and evil. The only word I could focus on was "bath." Here there was a problem. I had neither the key to my apartment nor any idea where it was. There was nothing for it but to go the main office in Fuld Hall and inquire. By this time Oppenheimer's classified files with their armed guard were long gone, and the outer office looked like any other. When I told the secretary who I was, and what I needed, she told me that Dr. Oppenheimer wanted to see me right away. This was so unlikely that I told her she must have me confused with someone else—probably someone of importance. No, she insisted, there was no confusion, Oppenheimer wanted to see me. When I walked into his office he was seated behind his desk, wearing one of those impeccable suits. The contrast between our general appearances is difficult to exaggerate. The first thing he said to me was, "What is new and firm in physics?" It was a two-part question. With luck I might have answered one part or the other. As I was trying to figure out what to say, the phone rang. I tried to excuse myself since I was sure that he wanted to answer the call in private. No, not at all.

When he hung up he said to me in a completely casual way, "It's Kitty," a reference to his wife. "She has been drinking again." I had absolutely no idea how to react to this, but before I could say, or do, anything he dismissed me with an invitation to drop by his house sometime because there were a few pictures I might like to see. Some weeks later a number of us were invited to a party there; the guests included the governor of New Jersey and John O'Hara, who was then living in Princeton. Very strong, very copious martinis were served, and it was only then that I realized that by the "pictures" Oppenheimer meant the van Gogh.

The visitors in physics my first year seemed to divide themselves into two groups. There were several German-speaking theorists, including Jost, who were trying to straighten out the formal aspects of the quantum theory of fields. These theorists were known as the "Feldverein," the field club. I liked them very much personally, but the mathematics was more than I could deal with. Once I was walking out of Fuld Hall after lunch when three of them, including Jost, were coming in the opposite direction. All of them were laughing very loudly. Jost had a laugh that was something like a braying animal. When they got close enough I asked them if they were laughing because they had just killed field theory once and for all. Jost replied in his broad Swiss accent, "We are laughing on the outside, but inside we are very sad." I was not sure how to interpret this.

Another large group was following the work of Lee and Yang. That is what interested me. I was in the process of retooling to work in this field, but I had gotten sidetracked because Oppenheimer had decided that we should all learn astronomy. Bengt Strömgren was persuaded to give a series of lectures, and we were persuaded to attend. In truth I found the first of these lectures incredibly dull; I now wish I had paid more attention, because astronomy has become such an exciting field, and one in which the Institute is playing a major role. In any event, I was daydreaming when I heard Oppenheimer call my name. He said I would have the job of preparing notes on the lectures. When I protested that I did not know anything about astronomy, he said, "Good, you will learn." Fortunately Strömgren was a kind and understanding man. Each week he would present me with a meticulous set of notes, which I would proofread and get duplicated.

This gave me the time I needed to retool. In the course of this I hit on something that actually interested both Lee and Yang and had a brief and very exciting collaboration with them. That spring, Bohr came to the Institute for one of his periodic visits, and a number of us were trotted out to present what we had been doing. Lee and Yang had much more important work to tell him about than our collaboration, so I got the job of speaking about it in one of our colloquiums at which both Bohr and Oppenheimer were in the audience. There were to be several speakers, and I was told to take no more than ten minutes. I took five. After my talk, Bohr said he thought it was very interesting. Only later did I realize that this was Bohrese for saying that it wasn't.

About this time I was called into Oppenheimer's office for my "confessional." It came at a time when I was tired of physics and was reading Proust, which I had never done in college. I was totally engrossed in the novel. When Oppenheimer asked me what I had been doing, I told him. He looked at me kindly and said that when he was about my age he had taken a walking trip on Corsica and had read Proust at night by flashlight. If I had known then what I know now, I might have asked him about the poisoned apple. Not long before the confessional we had sat together on a train ride to New York. He said a number of things that I have never forgotten. One of them was that he wanted to write a play which he was going to call *The Day That Roosevelt Died*. I think he still felt that had Roosevelt lived a little longer, we might have reached some sort of accommodation with the Russians on nuclear weapons. From what I have read subsequently in memoirs of people like Sakharov, I think this is an illusion. Once Stalin learned about Hiroshima, he was determined to have them at whatever price. On the train Oppenheimer told me that one of the first things he did when he became director of the Institute was to invite T. S. Eliot, one of his favorite poets, to visit. He went on to say that during his visit Eliot wrote his worst play, *The Cocktail Party*. Then he began talking about what he called "my case"—the hearing. He made the remark that while it was going on he felt as if it was happening to someone else. That was part of the tragedy. The hearing exposed the fragments of his personality. I have often wondered who the "someone else" could have been.

Apart from the tea lounge, whatever socializing there was at the Institute took place in the cafeteria. At night, gourmet dinners were served at very modest prices. People without families ate there regularly. As a rule, fields tended to flock together—physicists with physicists, mathematicians with mathematicians, archaeologists with archaeologists. The Institute had a strong group of archaeologists. One night I was sitting with my physics colleagues discussing an anomaly. The anomaly was that one of our well-known senior colleagues—known for his simple style of living and frugality—had been earning what must have been enormous sums consulting. What did he do with the money? The suggestion was made that he invested it. I noted that in this way he would only earn more money, which he wouldn't spend—enhancing the anomaly. At this point Dirac, who had been totally silent during dinner, spoke up. "Maybe he loses it," he said. We took the opportunity to ask him if he had ever collaborated with anyone in physics. "The really good ideas," he said, "are had only by one person." On another occasion I overheard a conversation between two mathematicians—a young and very aggressive American mathematician and a well-known senior French mathematician named Jean Leray. I had the distinct impression that Leray wanted to eat by himself but was being badgered into conversation. It went something like this:

M. Do you like movies, Professor Leray?

L. Silence.

M. What about gangster movies? Bang! Bang!

L. Silence.

M. Do you have gangsters in France, Professor Leray?

L. Yes, but they constitute the government.

Leray also figured in one of my favorite moments that first fall. We had gotten in the habit of playing touch football in the late afternoon on a sward in the back of the Institute. In due course we began to gather spectators. I think Oppenheimer watched once or twice. We made so much noise that we were asked to move farther away from the library. My team had on it several French mathematicians, including J. P. Serre of the Collége de France, who was considered to be one of the best young mathematicians in the world. The only play I could get my teammates to run consistently was *la statue de la liberté*. They liked the name. In any event, one of the mathematicians on the other team lost a contact lens, and the game stopped while we all began crawling around on the grass looking for it. At this moment Leray came by. He said to Serre, "What are you doing?" Serre replied, "We are playing American football, and are looking for the ball."

The cafeteria was also the setting of two remarkable encounters, one with Kurt Gödel and the other with W. H. Auden. The Gödel: For whatever reason, it had been decided that what we needed was a social mixer—what the Radcliffe girls used to call a "Jolly Up." This was to take place in the cafeteria one evening. There may even have been dancing, I forget. What I do remember is standing on the periphery and consuming several glasses of strong punch. Out of the corner of my eye I saw Oppenheimer and his wife escorting a somewhat frail-looking man with thick glasses. I immediately recognized him as Gödel. I had seen a great many pictures. He was one of my heroes. His theorem on the undecidability of mathematical truth was to me one of the most wonderful things I had ever learned and now, there he was, bearing down in my general direction. He was a notorious recluse, so it would never have occurred to me to have gone to his office. In any event, when the Oppenheimers stood within hailing distance they introduced me. Gödel then said he had known my father in Vienna. Since my father had never been in Vienna, I attempted to explain. Gödel then repeated that he had known my father in Vienna. I thanked him, and the Oppenheimers moved on.

The story with Auden is a little more complex. When I arrived at the Institute and looked at the roster of names, I noticed that of Reinhold Niebuhr. He was another of my heroes. I had heard him preach at Harvard and had read some of his books. He *was* a friend of my father's. One day in the cafeteria I introduced myself to the Niebuhrs—he was there with his wife, Ursula—and I received a cordial invitation to visit them in their Institute apartment. Ursula took me aside and explained that Niebuhr had recently suffered a stroke and would welcome the company of young people.

One evening the conversation came around to Auden, whom I had just heard read in Princeton. I think he was a poet in residence. Ursula immediately burst into enthusiastic whoops. She had been to college with Auden, and he had dedicated his poetry collection, *Nones*, to Niebuhr. I had more or less forgotten about this when on a train trip to New York—yet another—I found myself sitting next to Auden. I told him that the Niebuhrs were in Princeton, which he had not known. Some days later I got a call from Oppenheimer's secretary asking me to stop by the office at lunchtime. When I did, there were the Niebuhrs, Oppenheimer and Kitty, the historian Sir Llewellyn Wordward and his wife, and, of course, Auden. We all marched into the cafeteria for lunch. When my colleagues saw me, there were some fishy stares. What impressed me about that lunch was that no one had much to say. Oppenheimer told about studying Sanskrit and his Dirac story from Göttingen. None of the rest of us did much talking, and Auden looked bemused. After lunch I took him to meet Dyson, and they played word games on Dyson's blackboard.

What I did not know—Oppenheimer did not mention it—was that a year earlier Auden's contemporary, Stephen Spender, had also visited the Oppenheimers, or at least him. He had gone to their house but had not met Kitty, who had excused herself. It occurred to me that this might have had something to do with the Spanish Civil War, in which Spender, Kitty, and her then husband Joe Dallet had been involved. In any event, in his journals Spender describes his visit. He notes, "Oppenheimer lives in a beautiful house, the interior of which is painted almost entirely white. He has beautiful paintings. As soon as we came in, he said: 'Now is the time to look at the van Gogh.' We went into his sitting room and saw a very fine van Gogh of a sun above a field almost entirely enclosed in shadows." Spender then goes on to describe Oppenheimer's appearance. He writes, "Robert Oppenheimer is one of the most extraordinary-looking men I have ever seen. He has a head like that of a very small intelligent boy, with a long back to it, reminding one of those skulls which were specially elongated by the Egyptians. His skull gives an almost eggshell impression of fragility, and is supported by a very thin neck. His expression is radiant and at the same time ascetic." While this captures some of it, it leaves out the eyes, the luminescence of the blue eyes.

People who had known him before the hearing said that after it Oppenheimer was a changed—diminished—man. I did not know him from before, so I cannot say. But it is difficult to see how his personality could have been more powerful. I am sure that he was hit hard by the hearing, but I have often wondered whether a little of what people saw afterward was something Oppenheimer wanted them to see. I witnessed something that perhaps lends credence to this view. One Saturday morning I was in my office when a television camera crew began setting up in our seminar room. It was Howard K. Smith, the news commentator, and he was to do an interview. When Oppenheimer joined him and saw me, he said I could stay so long as I was in a place where he couldn't see me. A very odd request. Then the interview began, and Oppenheimer spoke to Smith in that low voice he used when he was talking about himself in this public way. When the camera stopped, the voice changed. He and Smith discussed the best French restaurants in New York and sailing in the Caribbean. The Oppenheimers now had a place on St. John's in the Virgin Islands. The days of Perro Caliente were long gone.

Both before and after the hearing, Oppenheimer was a public lecturer much in demand. In 1953, he gave the Reith lectures on the BBC. They were put into a book, *Science and the Common Understanding*. In the third lecture he spoke about the invention of the quantum theory. He said, "For those who participated it was a time of creation; there was terror as well as exaltation in their insight. It will probably not be recorded very completely as history. As history, its re-creation would call for an art as high as the story of Oedipus or the story of Cromwell, yet in a realm of action so remote from our common experience that it is unlikely to be known to any poet or any historian." Beautiful language, certainly, but is any of it true? Does it apply, for example, to Dirac? Sometime after the hearing, Oppenheimer gave a lecture at the University of Wisconsin. Someone who was there told me that afterward a woman asked a question. In reply Oppenheimer told sort of a joke, a putdown for asking the question which Oppenheimer must have considered stupid. A woman has a nightmare in which she wakes up to find a burglar at the end of her bed. She screams, "What do you want?" The burglar replies, "I don't know, lady, it's your nightmare."

Once I left the Institute I rarely saw Oppenheimer. I had heard rumors that with the arrival of the Kennedy administration, which contained old friends of his such as Arthur Schlesinger, Jr., McGeorge Bundy, and Dean Rusk, there was some reaching out. Oppenheimer never sought to have his clearance restored, and when asked about his advice on nuclear weapons he always said that he was too far removed from the subject to offer any. But in 1963 he was presented with the Fermi award, the most prestigious that the Atomic Energy Commission gave. It included a fifty-thousand-dollar stipend. A year earlier it had been given to Edward Teller. As was the custom of the committee that decided on the award, the previous winners made recommendations, and Teller nominated Oppenheimer.

It is hard to know what to make of this. When the award was announced in April 1963, Oppenheimer issued a statement which said, "Most of us look to the good opinion of our colleagues, and to the goodwill and confidence of our government. I am no exception." Before Kennedy could actually present the award he was shot—indeed, on the very day he had announced he would present the award himself. President Johnson decided to go ahead with the ceremony even though some who remembered his left-wing past opposed giving Oppenheimer this award. When he was presented the citation on December 2, Oppenheimer said to President Johnson, "I think it is just possible, Mr. President, that it has taken some charity and some courage for you to make this award today. That would seem a good augury for all our futures." There is a photograph of Oppenheimer and Teller shaking hands. The smiles on their faces seem genuine. So does the frozen look on Kitty's.

THE LAST TIME I saw Oppenheimer was also on the occasion of a lecture, though it was not one I attended. In the spring of 1963 he had come to New York to give a lecture that he called "The Added Cubit." Before the lecture he had gone to Columbia University, perhaps to see Rabi. While there he asked everyone what the title meant—where it came from. No one knew. As it happened, one of my Columbia colleagues called me with this story. I had no idea either, so I called the late Robert Merton who I was sure knew everything. He immediately identified it as coming from the New Testament—Matthew 6.27, "And which of you being anxious can add one cubit to his span of life?" What happened then is almost beyond belief. I went to midtown Manhattan to the Hotel Algonquin to meet some New Yorker colleagues. As I was passing the elevator, out walked the Oppenheimers. When he saw me he said, "Your father is a rabbi—you should know this." He had the wrong testament for my father, but I gave Merton's answer with no explanation. He looked at me very strangely.

I do not know whether at that time Oppenheimer already had the throat cancer that would kill him four years later. I do know that in those last years he acted with great courage. Even at the end he came to meetings of the physicists who were deciding who would be the new members for the following year, members he would not live to see. In 1965 he resigned as director of the Institute because his health would no longer permit him to carry out the work. But he remained on as a senior professor of physics. He died on February 18, 1967. For a few years Serber, whose own wife had died, and who had become Kitty's companion, organized small meetings at the Institute more or less in Oppenheimer's name. Kitty welcomed all of us. In 1972, Serber took a sabbatical from Columbia with the intention of making a transpacific voyage on Kitty's new boat, the *Moonraker*. She became ill suddenly and died in a hospital in Panama City. Serber himself died in 1997. As for the van Gogh, which he had painted in Saint Rémy in December 1889, it was sold by Kitty after her husband's death. It is now in the hands of a private collector who purchased it at auction at Sotheby's on April 24, 1985, for nine million dollars.

Oppenheimer accepting an honorary degree from Princeton, 1966

EPILOGUE

We waited until the blast had passed, walked out of the shelter, and then it was extremely solemn. We knew the world would not be the same. A few people laughed, a few people cried. Most people were silent. I remembered the line from the Hindu scripture, the *Bhagavad Gita*: Vishnu is trying to persuade the prince that he should do his duty, and to impress him he takes his multiarmed form and says, "Now I have become Death, the destroyer of worlds." I suppose we all thought that, one way or another.

— J. Robert Oppenheimer

I RECALL THAT February 25, 1967, in Princeton, was a bitterly cold day. This was partly the weather and partly the occasion. The occasion was a memorial service for Oppenheimer, who had died a week earlier. He had been cremated, and eventually his ashes were flown to the Virgin Islands and scattered in the ocean. Many people who had been associated with Oppenheimer had received formal invitations. Many came, including General Groves, Rabi, and Serber. I have kept the invitation card, but, unfortunately I did not keep the printed sheet that was handed out at the service. It had more details. There were three speakers: George Kennan, Hans Bethe, and Henry DeWolf Smyth. Bethe and Kennan had defended Oppenheimer at the hearing. Smyth was the Princeton physicist who wrote the report on the atomic bomb project that bears his name. He was also the only Atomic Energy commissioner who had voted to restore Oppenheimer's clearance.

One common theme of the talks was how much the hearing had affected Oppenheimer. Rabi once told me that it had nearly killed him. He had given his best, and, rather than gratitude, he had been publicly humiliated. In addition to the talks there was music. The Juilliard String Quartet performed two movements of Beethoven's Quartet in C Sharp Minor. There was also a somewhat unexpected recording of Stravinsky's Requiem Canticles. Upon looking at the printed program I saw that George Balanchine had selected the music, which presumably explains the choice of Stravinsky. I did not know it at the time, but Balanchine was seated at the service with the Oppenheimer family. He must have been present at the reception afterward hosted both by Kitty and Frank Oppenheimer.

As it happened, Balanchine and I attended the same Pilates gymnasium in New York, and I saw him there the Monday after the service. I told him I did not realize he was a friend of Oppenheimer's. He told me he had never met Oppenheimer and was surprised when he was asked to select the music. When I was thinking of writing this memoir, it occurred to me try to find out how this had happened. Balanchine was dead, as were the members of the

Oppenheimer family who might have remembered. None of the people who were at the Institute at that time remembered either. But finally I figured it out.

In the 1950s something called the Congress of Cultural Freedom was formed to provide a liberal, somewhat left-wing answer to Communist ideology. Oppenheimer joined as did many others, including Nicolas Nabokov, the novelist's cousin. In 1959 the Congress met near Basel, and Oppenheimer spoke. Among other things, he told the delegates, "I find myself profoundly in anguish over the fact that no ethical discourse of any nobility or weight has been addressed to the problem of the new weapons, of the atomic weapons . . . what are we to make of a civilization which has always regarded ethics as an essential part of human life, and which always had an articulate, deep, fervent conviction, never perhaps held by the majority, but never absent: a dedication to 'ahisma,' the Sanskrit word that means 'doing no harm or hurt,' which you will find in the teachings of Jesus and Socrates—what are we to think of such a civilization which has not been able to talk about the prospect of killing almost everybody except in prudential and game-theoretic terms?" Because of the Congress, Oppenheimer and Nicolas Nabokov became friends. People remember his being present in Oppenheimer's home as he was dying. Nabokov was then living in Princeton. He was working with Balanchine on a ballet based on Nabokov's music. He must have approached Balanchine. I have no idea if they discussed this with Oppenheimer. But there was a final irony. Neither Oppenheimer, Nabokov, or Balanchine knew that the Congress of Cultural Freedom had been secretly funded by the CIA. What would Oppenheimer and his enemies have made of that?

ACKNOWLEDGMENTS

HAVING WRITTEN a great many of them myself, I tend to distrust pages of acknowledgments. I usually assume that if the author were really honest, the acknowledgments would be much more interesting. Having said this, I can assure the reader that these acknowledgments are the genuine article. The reason why this help was especially necessary was the length of time it took me to write the book—about half a century. From the time I was an undergraduate at Harvard in the late 1940s, I began to collect anecdotes about my teachers and colleagues. Not because at the time I intended to write about them, but because the human side of things always interested me as much as the physics. The problem is that with the passage of time one misremembers, and that is why being reminded of what was actually true is so important. In addition, this subject is so complicated that no single individual knows it all. These acknowledgments represent the collective knowledge.

ACKNOWLEDGMENTS

In this spirit I thank Michael and Alice Arlen, Stephen Bernstein, Hans Bethe, David Brooks, Bengt Carlson, David Cassidy (who is writing his own fine biography of Oppenheimer), Robert Christy, Pat Crow, Gene Dannen, Freeman Dyson (whose skeptical observations are always interesting), Norman Dombey, Richard Garwin, Murph and Mildred Goldberger, David Griffiths, Gerald Holton, Edith Jenkins, Sara Lippincott, Roger A. Meade, Phil Morrison, Melba Phillips, Oliver Sachs, Arthur Schlesinger, Jr., Fritz Stern, Roger H. Stuewer, Carey Sublette (who knows more about the general subject of nuclear weapons than anyone), Tom Tombrello, Françoise Ulam, Ruth Vinson, Sallie Watkins, Morton White, and Bill Whitworth. I am saving a special thank you for Martin Sherwin, who knows as much about Oppenheimer as anyone can, and who read the manuscript line by line. I would like to hold these people responsible for any mistakes found in the book, but, alas, the mistakes are mine. Finally I thank Ivan Dee, who took on this project with skill and cheer. Many of the people I would like to thank are no longer here, but in memory, thank you.

NOTES

page

3   "I asked I. I. Rabi": Jeremy Bernstein, "Physicist," *The New Yorker*, October 13, 1975.

4   "Ehrenfest's certainty": Alice Kimball Smith and Charles Weiner, eds., *Robert Oppenheimer: Letters and Reflections* (Stanford, 1980), p. 121, hereafter referred to as Smith.

5   "an unsuccessful businessman": Smith, p. 2.

6   I am grateful to the van Gogh expert David Brooks for information about these paintings.

6   It can be obtained from the New York City Department of Records and Information Services, Municipal Archives, by mail or in person. I am grateful for the kind help I received from the personnel at this office.

7   "She was a very delicate person": Smith, p. 2.

7   "You must remember": Edith A. Jenkins, *Against a Field Sinister: Memoirs and Stories* (San Francisco, 1991), p. 22, hereafter referred to as Jenkins. I am grateful to Ms. Jenkins for several communications.

7   "It was nothing": Smith, p. 3.

9   The marriage certificate, which was supplied to me by the New York City Department of Records and Information Services, says that the marriage itself was performed at the Ethical Culture Society building.

9 "We propose": This can be found on the website http://www.acu.org/ adler.html.

10 "Several people": A recent example can be found in S. S. Schweber, *In the Shadow of the Bomb* (Princeton, 2000), hereafter referred to as Schweber 2000. He writes, "I have presented the ethos of the Ethical Culture Society in some detail because it molded Oppenheimer's early intellectual development and left, I believe, a deep imprint on him, particularly on his moral outlook" (p. 53). It is a theme of this book that Oppenheimer's "moral outlook" was deeply ambiguous and hardly reflects the ethos of the Ethical Culture Society.

10 "I was an unctuous": Denise Royal, *The Story of J. Robert Oppenheimer* (New York, 1969), p. 15.

11 "From conversations with him": Bernstein, "Physicist."

11 "I think the most": Smith, p. 4.

15 "I begin to believe": Smith, p. 54.

16 "Mr. Klock told me": Smith, p. 95.

19 "His problems": Smith, p. 77.

20 "My regret": Smith, p. 92.

25 "Oppenheimer caused me": Max Born, *My Life: Recollections of a Nobel Laureate* (Boston, 1978), p. 229.

27 "It was evening": Smith, p. 110.

31 "Prof. Takes Girl": I am grateful to Roger Stuewer for supplying me with three newspaper accounts of this event. The names of the papers and the precise dates are not given. I am also grateful to Melba Phillips, who was ninety-six when I interviewed her. I thank Ellen Vinson and Sallie Watkins for helping me arrange this interview and for additional information about Miss Phillips.

32 "Note on the Transmutation Function": The full reference is "Note on the Transformation Function for Deuterons" by J. R. Oppenheimer and M. Phillips, *Physical Review*, vol. 43 (1935).

35 "I am sending": Smith, p. 193.

36 "My friends": *In the Matter of J. Robert Oppenheimer: Transcript of Hearing Before Personnel Security Board and Texts of Principal Documents and Letters* (Cambridge, Mass., 1971), p. 8. This version has a foreword by Philip M. Stern, hereafter referred to as *Matter*. The original version, to which this is essentially identical, except that two documents have been combined into one, was issued by the Atomic Energy Commission in 1954, just after the hearing.

Both versions are now out of print and not easy to find. I would suggest avoiding the abridged versions which only hint at the full transcript.

37 "delightful memoir": The full reference is Robert Serber with Robert R. Crease, *Peace and War: Reminiscences of a Life on the Frontiers of Science* (New York, 1988), hereafter referred to as Serber 1998.

37 footnote information from *Matter*, p. 184.

38 "But it's such easy Dutch": This anecdote can also be found in Smith, p. 149, where the student is identified as Leo Nedelsky. Actually, I understand what Oppenheimer meant. When I was a graduate student, we were required to pass a language examination in German, a language I had never studied. But we were asked to translate a standard physics text with a lot of equations. You could make up the translation from the equations.

38 "his appreciation": Leonard Schiff, *Quantum Mechanics* (New York, 1955), p. vi.

39 "disappointed with Schwinger": For details, see S. S. Schweber, *QED and the Men Who Made It* (Princeton, 1994), hereafter referred to as Schweber 1994.

40 "Oppenheimer was a dominating": Schweber 1994, p. 288.

40 "Starting with": *Matter*, p. 288.

41 "His brother reported": This interview and several others can be seen in the 1980 John Else documentary *The Day After Trinity*. This superb film can be purchased for home viewing.

41 "In California": These anecdotes can be found in Smith.

42 "July 1945": Some of this footage can be seen in the Else documentary. It also shows the unmarked car that brought the plutonium for the test from Los Alamos. Philip Morrison accompanied the plutonium, and I am grateful for his comments.

47 "it seems unlikely": J. R. Oppenheimer and G. M. Volkoff, *Physical Review*, vol. 55 (1939), p. 380.

48 "On Continued Gravitational": J. R. Oppenheimer and H. Snyder, *Physical Review*, vol. 56 (1939), p. 455.

48 "in their group": Serber 1998, p. 48.

52 "Dear Ernest": Smith, p. 145.

52 "I am the loneliest man": Smith, p. 145.

53 "Beginning in late 1936": *Matter*, p. 8.

54 "She was private, too": Jenkins, p. 22.

54 "Never since the Greek tragedies": Jenkins, p. 23.

55   "she disappeared": Goodchild, *Oppenheimer*, p. 35.

55   "Jean stayed with us": Jenkins, p. 30.

56   "Between 1939 and 1944": *Matter*, pp. 153–154. Some of the chronological details can be found in Herken, *Brotherhood*, p. 102.

59   "For all the confidential exchange": Jenkins, p. 31.

60   "Jean was discovered": Jenkins, p. 38.

60   "I can remember": Haakon Chevalier, *Oppenheimer: The Story of a Friendship* (New York, 1965), p. 11.

61   "half jocular overstatement": *Matter*, p. 159.

62   "You know Oppenheimer": recorded interview with Rabi, undated.

62   "Pair Theory": *Physical Review*, vol. 61 (1942).

65   "We were lying": Frank Oppenheimer gives this description in the Else documentary. He told me the same thing.

65   "Stirling Colgate": I am grateful to Stirling Colgate for telling me about this. He also told me that, contrary to what one sometimes reads, the school was in good financial shape, so that its sale to the government was unrelated to financial exigencies.

68   "Memorandum on": These two reports may be found in Robert Serber, *The Los Alamos Primer* (Los Angeles, 1992), Appendix 1. This book is the text of the lectures that Serber gave at Los Alamos in the spring of 1943 to bring new recruits up to speed. For this edition he added much useful commentary. For someone with a technical background, this is still the best source for the physics of nuclear fission weapons.

69   "their program lasted": For readers who want more details supporting this claim, see Jeremy Bernstein, *Hitler's Uranium Club* (New York, 2001).

69   "Frisch and Peierls": For a more detailed account, see Richard Rhodes, *The Making of the Atomic Bomb* (New York, 1988).

70   "explosive chain reaction": For details about this and all the technical aspects of the Los Alamos program, see Lillian Hoddeson, Paul W. Henriksen, and Roger A. Meade, *Critical Assembly: The Technical History of Los Alamos During the Oppenheimer Years, 1943–1945* (New York, 1993).

71   "The day after": Serber 1998, p. 68.

72   "We had a compartment": This is based on interviews with Bethe that may be found in part in Jeremy Bernstein, *Hans Bethe: Prophet of Energy* (New York, 1980), pp. 72–73.

75   "Oh, that thing": This quotation and much on Groves can be found in

Robert S. Norris, *Racing for the Bomb* (South Royalton, Vt., 2002). The quotation is on p. 125.

76   "Dr. Oppenheimer was used": *Matter*, p. 164.

79   "In accordance with": *Matter*, p. 170.

82   "I would like to say": *Matter*, p. 712.

83   "When he told me": Edward Teller with Judith Schoolery, *Memoirs: A Twentieth-Century Journey in Science and Politics* (Cambridge, Mass., 2001), p. 177, hereafter referred to as Teller.

83   "Fortunately for me": Teller, pp. 177–178.

87   "I did suggest it": Smith, p. 290.

88   "In considering": Smith, pp. 316–317.

89   "Dear Haakon": In the Else documentary, Chevalier reads this letter.

93   "THE WITNESS": *Matter*, p. 468. I have maintained the punctuation of the original. The asterisks in the text represent an elision of classified material.

95   "DEAR DR. OPPENHEIMER": *Matter*, p. 3.

96   "But I was able": I am grateful to Hans Bethe for a telephone interview.

97   "Franck Report": For a discussion, see Nina Byers, "Physicists and the Decision to Drop the Bomb," *CERN Courier*, November 2002.

98   "He simply did not believe": Herbert York, *The Advisors: Oppenheimer, Teller and the Superbomb* (San Francisco, 1976), p. 34.

98   "If the United States": Byers, "Physicists."

99   "In fact he participated": For a discussion, see Martin Sherwin, *A World Destroyed: Hiroshima and the Origin of the Cold War* (New York, 1975).

100   "Oppenheimer and I": Bernstein, "Physicist."

102   "Hesitant and cheerless": Dean Acheson, *Present at the Creation* (New York, 1969), p. 151.

102   "Skobeltsyn," Rabi said: Bernstein, "Physicist."

103   "*Jump Book*": (New York, 1959), p. 8.

104   "In the Princeton": Halsman, *Jump Book*, p. 26. The Oppenheimer photograph is on p. 57.

105   "Strauss": For an account of Strauss's life, see Richard Pfau, *No Sacrifice Too Great* (Charlottesville, Va., 1984).

106   "The Commission is": The text of the full act can be found on the website www.nuclearfiles.org/redocuments/1945/461220-mcmahon.html.

108   "No one can force me": Stern, *Oppenheimer Case*, p. 129.

109 "Well, Joe": Stern, *Oppenheimer Case*, p. 13.

110 "It was very informal": *Matter*, p. 374.

111 "Beyond that": *Matter*, p. 374.

111 "Let me": *Matter*, p. 375.

113 "We are exploring": *Matter*, pp. 242–243.

115 "It was a real crisis time": Bernstein, "Physicist."

117 "We base our": York, *Advisors*, pp. 156–157.

119 "Necessarily such": York, *Advisors*, p. 158.

119 "In any event": Bernstein, "Physicist."

120 "Accordingly I": York, *Advisors*, p. 69.

121 "Now around this very time": Bernstein, "Physicist."

121 "It is my judgment": *Matter*, p. 81.

123 "are gone into": Especially useful to me was Carey Sublette's "Nuclear Weapons Frequently Asked Questions," http://www.nuketesting.enviroweb.org/hew/hew, where the diagram showing the Ulam-Teller design can be found. Also very useful was Bengt Carlson, "How Ulam Set the Stage," *Bulletin of the Atomic Scientists*, July/August 2003. I would like to thank Carey Sublette and Bengt Carlson for useful comments and Gene Dannen for calling my attention to Carlson's article. I am also grateful to Richard Garwin and Françoise Ulam for comments and corrections.

127 "Peters's story": It is well described in Schweber 2000.

129 "George and Dolly": Chevalier, *Oppenheimer*, pp. 52–55.

132 "He also leaves out": For more details, see Herken, *Brotherhood*, Chapter 3.

133 "The essence of the situation": Herken makes this an important theme of his book. A reader will have to decide on the plausibility of his arguments.

133 "letter to Oppenheimer": For the full text, see http//www.brotherhood-ofthebomb.com/hhbsource/document1.html.

134 "Whatever time you choose": *Matter*, p. 845. This is actually a second and corrected version that appears in the transcript. At the time of the hearing these tapes still existed. One wonders what happened to them.

135 "Well, I might say": *Matter*, pp. 845–846.

137 "to General Groves": Stern is a good source for this and other details. It also comes up in *Matter* but requires digging to ferret out.

137 "How about Chevalier?": Stern, *Oppenheimer Case*, p. 61.

138 "Q. Did you tell": *Matter*, p. 137.

139 "was Eltenton": See Herken, *Brotherhood*, pp. 54–55. For details.

140 "hydrogen-tritium": Alvarez testified extensively on this in the hearings. See *Matter*, p. 775 et seq.

140 "Richard Tolman": Herken, *Brotherhood*, p. 290.

140 "a broken heart": Herken, *Brotherhood*, endnote 1, p. 404.

141 "The precious times": I am grateful to Gregg Herken for sending me copies of these letters.

142 "Mr. Chairman": *Matter*, p. 680.

143 "a pattern of action": Wilson's testimony can be found in *Matter* starting on p. 679.

147 "This, in the opinion": This is taken from my copy of the hearing transcripts which, somehow, I got all the participants to sign, including McCarthy.

150 "In good sense": *Matter*, p. 1012.

151 "AEC's general counsel": Details may be found in Stern, *Oppenheimer Case*.

152 "the AEC's counsel": Pfau, *No Sacrifice*, p. 162.

152 "This I cannot do": Stern, *Oppenheimer Case*, p. 233.

153 "experienced litigator": Stern persuaded Garrison to write an extensive response to questions about the case. It appears in his book, pp. 504 et seq.

157 "Yes, I have known Communists": *Matter*, p. 671.

157 "In 1937": *Matter*, p. 8.

158 "My brother": *Matter*, p. 101.

159 "A whole chapter": Teller, Chapter 30.

159 "I was certain": Teller, pp. 373–374.

160 "At my request": Stern, *Oppenheimer Case*, p. 516.

161 "I tried to persuade": Bernstein, *Hans Bethe*, p. 99.

161 "Should Oppenheimer": Teller, pp. 374–375.

162 "Q. To simplify": *Matter*, p. 710.

163 "But my most glaring error": Teller, p. 383.

163 "MR. GRAY": *Matter*, p. 726.

165 "Of course": *Matter*, p. 1010.

166 "Most of the derogatory information": *Matter*, p. 1020.

166 "Q. Doctor": *Matter*, p. 723.

168 "In my opinion": *Matter*, p. 1043. The asterisks are unexplained in the original.

169 "The American Government": Goodchild, *Oppenheimer*, p. 272.

178 "A *Philosopher's Story*": Philadelphia, 1999.

178 "Robert is never wrong!": White, *Philosopher's Story*, p. 140.

179 "He is a second Dirac": Smith, p. 269.

180 "We have been at the ranch": Smith, pp. 301–302.

181 "I came in the late summer": *Matter*, pp. 26–27.

182 "relative to Julius": Goodchild, *Oppenheimer*, p. 183.

182 "offering a seminar": This is described in detail in Freeman Dyson, *Disturbing the Universe* (New York, 1979).

193 "Oppenheimer lives in a beautiful house": Stephen Spender, *Journals 1939–1983*, ed. by John Goldsmith (New York, 1986), p. 182.

194 "For those who participated": J. Robert Oppenheimer, *Science and the Common Understanding* (New York, 1954), pp. 35–36.

195 "Someone who was there": I would like to thank my brother, who was there, for telling me this anecdote.

195 "Most of us": This and the next quotation come from Goodchild, *Oppenheimer*, p. 275. Some measure of the opposition that Oppenheimer still engendered was reflected in the fact that when my unsigned obituary of him appeared in *The New Yorker*, William Shawn told me he received objections from some people who claimed that Oppenheimer was a Communist.

197 "It is now": I would like to thank the van Gogh expert David Brooks for this information.

199 "We waited": These are the last words Oppenheimer says in the John Else documentary.

201 "Congress of Cultural Freedom": A history of this institution can be found in Peter Coleman, *The Liberal Conspiracy* (New York, 1989). The quote that follows is on p. 121.

INDEX

A NOTE ON THE AUTHOR

Jeremy Bernstein was born in Rochester, New York, and educated at Harvard University. He is a theoretical physicist who is now professor emeritus of physics at the Stevens Institute of Technology. He has held appointments at the Institute for Advanced Study at Princeton, the Brookhaven National Laboratory, CERN (the European Organization for Nuclear Research), Oxford University, the University of Islamabad, and the École Polytechnique.

From 1961 to 1993 Mr. Bernstein was a staff writer for the *New Yorker*, where he published a great many articles on science and scientists. His books include *Einstein* (nominated for a National Book Award); *Three Degrees Above Zero*; *Mountain Passages*; *Science Observed*; *The Tenth Dimension*; *Cosmological Constants*; *Quantum Profiles*; *Cracks, Quarks, and the Cosmos*; *Hitler's Uranium Club*; and *The Merely Personal*. Among a great many honors and awards, he has won writing prizes from the American Association for the Advancement of Science and the American Institute of Physics. He lives in New York City and Aspen, Colorado.